SAVING WATER
IN A
DESERT
CITY

SAVING WATER IN A DESERT CITY

William E. Martin
Helen M. Ingram
Nancy K. Laney
Adrian H. Griffin

RESOURCES FOR THE FUTURE / WASHINGTON, D.C.

Resources for the Future books are distributed worldwide by The Johns Hopkins University Press.

Manufactured in the United States of America

Published May 1984

Library of Congress Cataloging in Publication Data
Main entry under title:

Saving water in a desert city.

  Bibliography: p.
  Includes index.
  1. Tucson (Ariz.)—Water-supply—Rates. 2. Water conservation—Arizona—Tucson. I. Martin, William Edwin. II. Resources for the Future.
HD4464.T8S2 1984   333.91′216′0979177   83-43263
ISBN 0-915707-04-7

Contents

Tables

Figures

FOREWORD

One facet of RFF's continuing interest in water resources has been concern for municipal water system problems. This concern has been expressed in studies of residential water demand (Howe and Linaweaver, "The Impact of Price on Residential Water Demand . . .," *Water Resources Research* vol. 1, 1965); of actual and economically optimal methods of planning and building reservoirs to cope with drought (Russell, Arey, and Kates, *Drought and Water Supply*, RFF, 1970); of drinking water safety and the costs and benefits of the Safe Drinking Water Act (*Safe Drinking Water*, Russell, ed., RFF, 1978); and of groundwater contamination from hazardous waste disposal sites (Sharefkin, Schechter, and Kneese, "Impacts, Costs, and Techniques for Mitigation of Contaminated Groundwater," in National Science Foundation, *Workshop on Groundwater Resources and Contamination in the United States*, NSF PRA Report, 1983). These studies, while including necessary elements of engineering and science, concentrated on economic problems—of demand estimation, system expansion, and benefit estimation.

This book was stimulated by and sets out to analyze a *political* battle over water pricing by a municipal system. In the course of that analysis it provides us with improved methods for demand function estimation where block rates are involved, suggests procedures for rational pricing of municipal water, and is otherwise faithful to RFF's economics tradition while attempting to explain how politics can dominate when real decisions are made. It also shows how the emotionally powerful water use and development issues can stand in for economically larger concerns like the pace and character of urban growth and change.

Therefore, *Saving Water in a Desert City* may be said both to continue a tradition and to break some new ground for RFF's publications about municipal water problems. Because it has the additional virtue of being easy to read, it should be of wide interest to those involved with municipal services and resource management in general.

Clifford S. Russell

December 1983   *Washington, D.C.*

Acknowledgments

While this is a book about Tucson, Arizona, written by Tucsonans, the lessons of the Tucson experience with water conservation would have been lost to others, and perhaps even to the city itself, had it not been for crucial resources contributed from outside. Water conservation was a national water policy objective of the Carter Administration in the late 1970s and a research funding priority. One of the authors was at that time a visiting scholar at Resources for the Future and was encouraged by fellow researchers to see the relevance of hometown events for addressing the national questions of why, how, and when municipal water supplies should be conserved. The research which undergirds this book was initiated at Resources for the Future and carried out mainly at the University of Arizona.

The authors gratefully acknowledge the Office of Water Resources and Technology, U.S. Department of Interior, (later the Office of Water Policy), particularly Dr. Madge Ertel, the project manager, for partial financial support. Further, the authors are indebted to Resources for the Future both for stimulating the research effort and applying its usual rigorous analytical and editorial review and advice to improving this product.

1

Conundrums About Water Scarcity

The message about resource scarcities reaching the American public seems clear enough on its face. Shortages of vital natural resources are proclaimed and predicted in no uncertain terms by the media. The oil crisis, the energy crisis, food shortages, dwindling farmland, and shrinking forests all capture the periodic attention of the media and hence of the public. Scarcity is a theme that has become an important part of the rhetoric of recent times. Yet words often are contradicted by actions. There is a curious gap between the strident warnings that scarcity is imminent and effective policies to address presumed shortages.

The most vivid and recurring issue in resource scarcity has been that of dwindling water supplies. "The Water Crisis: It's Almost Here," proclaims an article in one popular magazine, while another forecasts "The Browning of America" in its cover story.[1] A leading scientific journal poses the question of "What to Do When the Well Runs Dry" to introduce an article on the Ogallala Aquifer of the Plains states, a subject also discussed in a newsmagazine story subtitled "The Great Water Hole Beneath the Great Plains Is Going Dry."[2] *The Global 2000 Report* anticipated an increase of 200 to 300 percent in world water withdrawals over the 1975–2000 period, and estimated that regional water shortages and deteriorating water quality are likely to become worse.[3] Despite these alarming prognostications, most water management and pricing policies continue to reflect an era of abundant supplies, and in most places people continue to use water as if scarcity was not an issue.

The contradictions between messages about water scarcity and the management and use of the resource are particularly apparent in the arid Southwest. There, where green oases made possible by water development contrast sharply with the surrounding arid terrain, one might expect to find rational and orderly management of water resources. In areas where it is common to hear the phrase "Water is survival," there should be a great incentive to put one's beliefs about water scarcity into practice. However, instead of frugally managing their water supplies, southwestern desert cities generally may be characterized as engaging in business as usual. Urban water departments in the Southwest, as those in the more humid regions of the country, have looked primarily toward supplying all customers with as much inexpensive water as they might want for whatever purposes.[4] People have responded by using large quantities of water. The public may believe that water is scarce and should be conserved, but those beliefs have not yet been translated widely into public policy or private practice.

This book examines the recent history of municipal water management and use in the southwestern city of Tucson, Arizona, and the contradictions and ambiguities of perceptions and policies that have affected that history. In summing up the Tucson experience, we argue that managers and customers of municipal water supply systems could well learn from the example of Tucson if they are to avoid the hardships of making similar discoveries for themselves.

Tucson, Arizona, represents what may be a prototypical example of the conflicts between beliefs and practices in the area of water use. A desert city with a population of nearly 0.5 million in 1983, Tucson and its surrounding area are expected to have 1.8 million inhabitants in 2035.[5] Table 1-1 indicates the enormous past and possible future growth in water demand in Tucson. Currently there are no developed surface-water supplies. Entirely dependent on groundwater, the City of Tucson pumps from depths of up to 450 feet, with the pumping depth

Table 1-1. Actual and Projected Growth in Demand for Water from the City of Tucson Water Department, 1965–2000

Year	Service population[a]	Average daily water demand (million gal/day)[b]
1965	232,000	41.5
1970	289,000	54.1
1975	380,000	67.6
1980	439,000	69.7
1985	494,000	81.8
1990	599,000	99.2
1995	685,000	113.4
2000	823,000	176.3

Source: City of Tucson Water Department, 1982.
[a]Service population is based on services × 3.7 until 1975 and 3.5 thereafter.
[b]Future estimates are based on the 165.6 gallons per capita per day used during fiscal year 1981/82. Population projections are from the Arizona Office of Economic Development and the Tucson Water Department.

falling as much as 5 feet a year in some areas. Falling water tables are due to the fact that 76 percent of consumptive water use is obtained through overdraft, that is, pumpage in excess of recharge.[6,7]

With Tucson residents facing these physical facts, water has become an important local political issue. Even so, the city's approach to water management still is at best ambiguous. Water scarcity is considered to be a serious problem by most Tucsonans.[8] Politicians in Tucson are fond of saying that water is among the most vital public issues. Nevertheless, water policy is for the most part set not by elected officials, but by professionals in the City of Tucson Water Department, the city's municipal water utility, with the advice of a nonpolitical citizens' group. Through publicity campaigns the Water Department urges its customers to save scarce water, yet water rates are kept as low as possible, and rate increases have not in recent years kept up with inflation. Such pricing policies encourage water use. Despite fears that there is not enough water, hydrologists do not know and studies have not been funded to determine precisely how much water is in the aquifer that underlies Tucson. Instead, the city is actively seeking construction of a massive federal water project, the Central Arizona Project (CAP), which is to bring Colorado River water into Phoenix, Tucson, and the surrounding agricultural areas. (As of this writing, the Phoenix portion of the project is almost complete. Actual construction on the Tucson portion has not yet begun. There is a slight possibility that that leg of the project will not be funded.) Project water, if it ever reaches Tucson, will be very expensive and its quality will be much lower than that of existing groundwater supplies. At the most optimistic predicted level of water delivery, the project will supply only about half of the present 1983 groundwater overdraft.

But while the groundwater table is falling, there clearly is no current physical water shortage in the Tucson area. Approximately 64 percent of annual pumpage in the Tucson Basin, and the adjoining Avra Valley Basin from which the municipality pumps about 25 percent of its water supply, is devoted to irrigated agriculture.

If additional supplies of water were needed for Tucson, the city could buy out the farms and use the water itself. In fact, it has already begun this procedure in order to obtain the Avra Valley water that it currently uses.

The groundwater decline could be halted, at least for several decades, if all agricultural pumpage in the two basins was halted. However, there is no likelihood of such a measure. Further, even though an eventual water balance has been mandated by the Arizona Groundwater Management Act of 1980, there is no pressing physical or economic need to achieve this balance. The groundwater is there in large (but precisely unknown) quantities, and in many areas is still relatively close to the surface in economic terms. People argue about the urgency of measures needed to provide for future water supplies and to conserve water, yet urban development is encouraged. Everyone can agree that ultimately something must be done, especially if the population continues to grow as expected, but what must be done is still a matter of diverse opinion.

Tucson is not unique in this dilemma—ambiguity in beliefs about water and in management of water exists everywhere. To a considerable degree, the situation in Tucson foreshadows what will be a common problem among many cities in the future. Higher living standards and urbanization have resulted in ever-increasing demands for water for relatively small geographic areas. New reservoir sites are scarce, groundwater is often available only as a supplemental source, and available water may not be of acceptable quality. Interbasin transfers of water are economically and environmentally costly, and areas of origin often resist the transfer. Pumping costs become higher as pumping depths increase and energy costs rise. Higher interest rates are a real constraint on capital-intensive construction projects. Supply alternatives to water problems may not be viable solutions in many urban areas in the future.[9] Yet municipal water utilities have traditionally engaged in management practices that encourage growth in demand, and most continue to do so.

What does tend to set Tucson apart from other muncipalities is that beginning in fiscal year 1973/74 the city experienced a significant reduction in per capita water use. Between fiscal years 1973/74 and 1978/79, water pumpage declined from 205 gallons per capita per day to fewer than 150 gallons per capita per day. Because of this reduction in per capita use, Tucson has been cited widely as a model for successful conservation management. But the reasons for the reduction in water use in Tucson have not been explained, and success may not yet be achieved. A careful examination of what happened in Tucson may offer lessons for other cities interested in encouraging conservation behavior.

It is clear that the decline in per capita water use in Tucson did not come easily or without cost. Tucsonans reacted with anger and dismay to a change in traditional water policy. The gap between the physical facts or possible future scarcity and current behavioral practices in using water was firmly entrenched in Tucson as it is elsewhere. For Tucsonans, establishing new patterns of water use was, and continues to be, a traumatic experience from which some lessons in politics, economics, and conservation management may be gleaned.

Water Lore

Water is considered by many to be a fundamental resource, somewhat different from other natural resources. Water is basic to human activity, and the sustenance of life as well as the production of almost every good requires water. Water plays an important role in law and in many religions, and the secure supply of water is among the elementary tasks of community organization. Water resources scholars have remarked upon the strange, emotional, and seemingly irrational way in which people, particularly westerners, view water resources. Maurice Kelso remarks that instead of water's being treated like any other commodity, it is treated as if it is priceless, that is, as being so precious that it cannot be given a price and instead should be available practically free in whatever quantities a person can use. Water is viewed as the key to prosperity—it can make the deserts bloom and it can transform a dusty crossroads into a thriving metropolis.[10] Water has the legendary Midas touch, and the ripple of the stream is also the rich jingle of coins. Cutting back on water use arouses enormous opposition.

The ambiguity with which water is treated by cities may spring from the separate and partially conflicting roles water plays in society. It has symbolic and aesthetic values as well as values in production and human consumption. Kenneth Boulding has suggested that water allocation policy may never be improved because it is so affected with romantic and poetic interests that it cannot be dealt with rationally, as economists see rationality.[11] Yet water has a price and an economic value, and in the West water is being slowly reallocated away from low-value uses in irrigated agriculture to municipal and industrial uses. It is an old saying that water flows uphill toward money. At the same time, governments have never been satisfied with the allocation of water through markets. Studying irrigation communities, Arthur Maass and Raymond Anderson found that the goals of maintaining order and certainty and resolving conflicts over water use were of higher priority than was efficiency in water allocation.[12] The aims of municipal governments seem to follow similar priorities. Finally, because water is closely involved in everyday life, its use is heavily dependent upon circumstances and habits. It may be only in unusual or stressed circumstances that the separate human concerns about water begin to mesh and its treatment becomes less ambiguous.

Boulding hopes that a more rational attitude toward this substance may develop in an age of increasing scarcity of energy and materials. A practical and realistic approach to water allocation will evolve, according to Boulding, through the interaction of three P's—preachments, policemen, and prices.[13] We propose modifying "policemen" to the more inclusive and less coercive-sounding "politics," and adding a fourth P, practices, to encompass the habitual nature of water use. Perhaps by unraveling the determinates of water use and conservation we can provide insights into a more general resource conservation policy.

Preachments

To economists Kelso and Boulding, society's actions with regard to water appear irrational because they are poorly related to economic value, that is, to the market value of water in alternative uses. Water is part of the symbolic value system and as such is part of our moral order. Water is considered too important a commodity for society simply to allow its use to be directed by money. Because water is viewed as a part of every human's basic right, it is considered practically immoral to charge more than a very low price for water. Consequently water is priced too low in market terms, and as a result it is overused relative to its real cost and its economic (productive) value. Thus, because water tends to be allocated by rules rather than being sold at market prices, some people use water profligately or in low-productivity enterprises, while other people who could put the resource to more productive use find that water is not available. The problem is one of basic preferences, ethics, and ideology, all of which may be affected by preachments.

Changing the uses of water involves altering the preachments that orient society's decisions about how water is allocated and used. Among the relevant ideas governing water use, the concept of conservation holds a powerful position. Notions that resources should not be squandered and that waste is immoral are important norms in our society. Yet water conservation, except in times of drought or emergency, has not had much practical application. One of the objectives of this book is to explain why. We argue that the contradictions that arose in the historical development of the concept of conservation and the divorce of that concept from economics have crippled the implementation of conservation in municipal water systems.

Among the other relevant ideologies that affect water use and which might be altered through preachments are the professional perspectives and organizational doctrines of water utility managers. The prevailing objectives of water managers have been to supply constant, dependable, and cheap service to everyone without error or controversy. Running a water utility is a professional rather than a political affair. For water managers to allow the delivery of water to become a political issue is to fail in their professional duty. When all users are served with plentiful, cheap water, there is no basis for conflicts between different interests. Insofar as the concept of conservation implies any users being forced to cut back on use, the orientation of traditional water utility managers has been unsympathetic. Cutting back water use in municipal water supply systems involves changes in management thinking. Our case study of the Tucson municipal water system offers an example of how such a change may come about.

Prices

The price charged for a commodity has a great deal to do with how much of the commodity is used and who uses it. Almost without exception, the water rates charged by municipal water systems have not been designed to reflect that fact.[14] From an economist's perspective, water is priced appropriately when its price is equal to the water's marginal cost of production. Under these circumstances a buyer will purchase only that quantity of water for which its additional (marginal) benefits to the buyer exceed the additional (marginal) cost of supplying the water. Such a pricing scheme ensures an "efficient" allocation of the resource. When water is priced too low, too much is used, and the marginal cost of producing the water exceeds the value of its marginal use. In this case, an "inefficient" allocation of the resource occurs.[15] With little economic incentive to conserve water, it is not surprising that municipal users have used water in large quantities, resulting in inefficient use.

Pricing also should have a time perspective. If something plentiful and cheap now is going to be more scarce and expensive to produce later on, the sensible social policy is to make it expensive right away. It is important to anticipate scarcity with present prices; otherwise society will be ill-prepared when shortages arise.[16] Cheap prices now encourage current use, hastening the time of real scarcity. Few will conserve now when the signal of the low price says that the commodity is abundant.

Although economists' arguments about pricing seem compelling intellectually, they have had practically no influence on municipal water utility management. Water utility managers have not been interested in economists' concepts of efficiency, but only in raising enough capital to construct new waterworks as "needed." The concept of need is not defined, but basically is related to historical per capita use. Economists, for their part, have not attempted to communicate with utility managers in terms the managers can understand. Further, as we will show, economic thinking has been divorced from other influences promoting change, particularly conceptions of conservation. On the one hand, until recently most economists have refused to consider scarcity in terms other than relative market prices and the allocation of scarce resources so as to achieve economic efficiency. There has been little interest in equity or distribution. On the other hand, many conservationists have refused to accept economic efficiency as a criterion for resource use, because for them the issue was emotional and ethical.

It seems certain that effective long-term water conservation requires the use of price as an incentive to save. The experience of Tucson is testimony to the effectiveness of raising water prices, even though Tucson's pricing policies have not been exemplary. Economics must be integrated into the thinking of public utility managers and into the concept of conservation. The physical reality of growing water scarcity and the increasing financial and environmental costs of developing new sources may provide the opportunities to introduce more appropriate pricing strategies that take into account marginal costs of production. However, establishing the economic rationality of a rate structure is not at all the same as addressing political feasibility, which brings us to the third P among the determinants of changing water allocation—politics.

Politics

Politicians gain political support by fostering perceptions of benefits, particularly among the most active and influential citizens. Politicians, especially those who are elected, cannot survive if it is widely thought by influential people that they are worse off due to politicians' actions. Making something as widely used as water more expensive fosters perceptions of costs, not of benefits. Under ordinary circumstances, it is not in a politician's interest to raise water rates.

In fact, under ordinary circumstances politicians are unlikely to get involved in municipal water systems at all. Delivering water is considered a professional matter to be taken care of by experts hired by the city. If the water department becomes a public issue and an act of God or nature cannot be blamed, it is generally supposed that there is something the matter with the water department. In most cities water rate setting is institutionalized into the bureaucratic agenda where decisions are made by experts within agencies and follow closely the pattern of past decisions. Substantial changes in water allocation or pricing policy require elevating an issue to the political or policy agenda where political actors tied to electoral constituencies are involved. Ordinarily such an agenda change requires a dramatic situation such as extreme drought or scandal. This book explores some other incentives that may operate to involve elected politicians in water utility decision making and to open possibilities for nonincremental policy change.

A particular focus of our analysis will be how the political costs of doing something hurtful to water users may be handled. It may be possible to create countervailing perceptions of benefits that can help soften the political impact. To some extent, some politicians may simply have to bear the burden of blame; how can they be persuaded to do so? Conservation policies are likely to be authorized only when they are politically feasible. Further, they are effective only when they can be implemented, that is, only when they actually change water use practices.

Practices

Few actions are as mindless as turning on the water tap. In fact, crucial decisions about whether and when faucets are used are often made by consumers in contexts far removed from that of water use: decisions such as whether to purchase evaporative cooling rather than air conditioning, how many rooms and bathrooms a home should have, and what kind of landscaping is preferred. To raise the quantity of water use to the level of a conscious decision on the part of the water user requires some jolt or shock. Surprises relating to a commodity as elemental and emotion-laden as water supply are not likely to be well received.

Research and experience have shown that water users can cut use substantially in the face of a crisis such as drought if the crisis is perceived to be real and if people believe that the burden is being shared equitably.[17] Useful as these findings are, it is difficult to extrapolate to noncrisis circumstances and conditions in which saving strategies are not applied across the board. Evidence suggests that water use returns to normal once a short-term water crisis is over. This volume is concerned with continuous, long-term water-saving practices. Under what circumstances can users be persuaded to change their water-using habits, and what is the relative importance of such factors as weather, landscaping, price changes, fashion, peer pressure, and public information?

The Objective of Policy Analysis

Altering the pattern of water allocation and use requires an integration of the influences we have labeled the four P's—preachments, prices, politics, and practices. The task of satisfying four conditions at once is difficult, because what appears reasonable and rational in one context is impermissible in another. For instance, conservation as defined in the preachments of some environmentalists may have little relation to what economists identify as efficient resource use. Price increases that an economist would view as efficient may be infeasible when evaluated in a politician's balance of potential political costs versus benefits. What is perhaps politically feasible—for example, incremental adjustments in past pricing and public information policies—may not have perceptible impact upon the practices of water users whose water-using practices are habitual. The role of policy analysis, as the authors of this volume see it, is to identify those courses of action that simultaneously are most likely to satisfy the critical determinants.

We see policy analysis as a multidisciplinary activity, in this instance encompassing the fields of economics, political science, and resource management. As Aaron Wildavsky states, "An analysis of public policy that does not consider incompatibilities among the realms of rationality is bound to be partial and misleading."[18]

Where does the analyst look for policy models that integrate conflicting and partially incompatible strictures? Giandomenico Majone has argued that policy analysis is more a critical interpretation of the past than a prediction of the future. By imagining what might have been, it is possible for an analyst to prepare for an improved future. The analyst assesses the adequacy of arguments, the strength and fit of the evidence, the relevance and reliability of data, the intrinsic limitation of policy tools, and the pitfalls lurking behind every conclusion.[19] Following this logic, we should look for a design for integrating the four P's in a real-life situation where water savings have occurred. It is not important that the case be exemplary. It is possible to learn as much from error as from success. It is not even important where the integrated strategy was deliberate and consciously pursued. According to Wildavsky, rationality is like a rocker that goes forward and back; it tries by intention and is saved by rationalization. One acts first and makes sense of it later. We rewrite history from present motives. By attributing new motivational meaning to what we have done, we try to learn what we ought to be doing.[20] Herein lies the opportunity for case studies such as the Tucson experience.

The Tucson Case Study

In 1976 the City of Tucson Water Department faced what it perceived to be a crisis. The expansion of its service capacity and the development of new water sources had not kept pace with enormous population growth and urban sprawl.

New service connections, some in the foothills requiring expensive pumping costs for water delivery, were imposing financial burdens that could not be carried by the prevailing low water rates. Rather than incrementally adjusting its water rates, the utility proposed a sharp increase in charges to customers and a new rate schedule intended to reflect the costs of service. The Tucson City Council, governed by a newly elected majority favorably oriented toward controlling growth, collaborated in the imposition of the new rate increases. The resulting sharp rise in some customers' water bills spawned a recall campaign in which the councillors responsible for the rate increase were retired from office. The replacements did not initiate a return to previous rate levels, however, and despite the change in council majority, rates were increased even further. A public information program aimed at lowering peak watering on hot summer afternoons was instituted. Tucsonans responded by reducing their use of water by approximately 16 percent per capita and sharply curtailing peak uses.

Per capita water use had reached a maximum of 205 gallons per day in fiscal year 1973/74, and, in fact, apparently was already falling by the time of the 1976 political crisis. The total drop of what was thought to be approximately 30 percent in per capita daily consumption[21] has made the city's experience a popular subject for research on water conservation. In seeking explanation for the presumed success of water demand management in Tucson, observers mostly have credited the program of public education. The program may have been important, but this explanation has done little to unravel the dilemmas posed by the foregoing discussion of the four P's. The challenge for policy analysis is to rationalize, that is, to create a postdecision explanation of events and actions. This task of retrospection can serve to incorporate the lessons of the past into the present and help us design a future direction for Tucson as well as for other municipal water supply systems.

Design of This Study

While the art of policy analysis is not so much in telling the story as in blending together diverse and conflicting perspectives on the question at issue, it is important for analysts to share with each other and the reader some basic facts about problems, actors, actions, and consequences. Chapter 2 contains a narrative of the Tucson experience containing sufficient detail to support the analysis that follows. The Tucson case is purported to be an example of water conservation; the concepts of conservation and the confusions that have arisen about the alternative concepts are the subject of chapter 3. Since we argue that the Tucson case exemplifies the need for a new integrated conception of conservation that includes economics, chapters 4 and 5 contain an analysis of water pricing and the significance of price in Tucsonans' conservation behavior. In chapter 4, "Pricing Municipal Water," the first six sections are somewhat more technical than the rest of the book. Those sections cover the economics of utility rate making and give a detailed argument for marginal cost pricing. Some readers may prefer to begin with the section entitled "Tucson Utility Rate Schedules," and return to the first portion of the chapter only if they want more technical details.

The analysis of water demand in chapter 5 finds that price had a crucial effect on Tucsonans' water use behavior, and that the very difficulty of political action on prices reinforces consumer perception of prices. Thus, chapter 6 focuses on the politics of water conservation, including the political benefits and costs to politicians of pursuing a conservation strategy as well as the impact of politics on water users' practices. Chapter 7 examines the practices available to the individual water user toward whom conservation strategies are directed. The major determinants of water use are identified, and the degree of flexibility that exists is examined. The final chapter sums up the issues discussed, particularly the questions of whether and to what extent it is possible to change water management practices so that they correspond better to physical and economic realities. We argue that both managers and customers of municipal water supply systems could well learn from the example of Tucson and avoid the hardship of later facing similar problems themselves.

Notes

1. K. K. Wiegner, "The Water Crisis: It's Almost Here," *Forbes* (August 20, 1979) pp. 56–63; and Jerry Adler, W. J. Cook, Strylier McGuire, G. C. Lubenow, Martin Kasindorf, Frank Maier, and Holly Morris, "The Browning of America," *Newsweek* (February 23, 1981) pp. 25–37.

2. John Walsh, "What to Do When the Well Runs Dry," *Science* vol. 210 (1980) pp. 754–756; and R. Stengel, "The Ebbing of the Ogallala: The Great Water Hole Beneath the Great Plains Is Going Dry," *Time* May 10, 1982, pp. 98–99.

3. Gerald O. Barney, *The Global 2000 Report to the President, Entering the Twenty-First Century* (Middlesex, England, Penguin Books, 1982) p. 26.

4. Steve H. Hanke, "Pricing as a Conservation Tool: An Economist's Dream Come True?" in David Holtz and Scott Sebastian, eds., *Municipal Water Systems* (Bloomington, Indiana University Press, 1978) pp. 221–222.

5. The estimate of 1.8 million people was officially adopted by the Arizona Department of Economic Security, Office of Planning, on December 18, 1981, and is the planning figure used by the Tucson Water Department. The area for the population projection includes most of eastern Pima County and is much larger than the area within the city limits.

6. Arizona Water Commission, *State Water Plan, Phase I* (Phoenix, Ariz., 1975).

7. Pumpage is not the same thing as "consumptive use" (figures for the latter are reported in table 2-1 of chapter 2). Both agriculture and nonagriculture consume less than they pump, since some water is returned to the aquifer. Consumptive use is the more important concept relative to the overdraft on the groundwater supply, but pumpage gives a better idea of relative water-using activities and of the costs of obtaining the groundwater supplies for use.

8. Helen M. Ingram, Nancy K. Laney, and John R. McCain, *A Policy Approach to Political Representation: Lessons from the Four Corners States* (Baltimore, Md., Johns Hopkins University Press for Resources for the Future, 1980) pp. 109–111.

9. David Holtz and Scott Sebastian, eds., *Municipal Water Systems* (Bloomington, Indiana University Press, 1978) p. 3.

10. Maurice M. Kelso, "The Water-Is-Different Syndrome, or What Is Wrong with the Water Industry?" in *Proceedings of the Third Annual American Water Resources Association Conference* (Urbana, Ill., 1967).

11. Kenneth E. Boulding, "The Implications of Improved Water Allocation Policy," in *Western Water Resources: Coming Problems and the Policy Alternatives* (Boulder, Colo., Westview Press, 1980) p. 300.

12. Arthur Maass and Raymond L. Anderson, *And the Deserts Shall Rejoice: Conflict, Growth and Justice in Arid Environments* (Cambridge, Mass., MIT Press, 1978) p. 1.

13. Boulding, *"Improved Water Allocation Policy,"* p. 301.

14. Hanke, "Pricing as a Conservation Tool," pp. 221–222.

15. Ibid., p. 225.

16. Boulding, *Improved Water Allocation Policy*," p. 305. The theory of optimal depletion over time has a long history, of which one of the early rigorous expositions was Harold Hotelling, "The Economics of Exhaustible Resources," *Journal of Political Economy* vol. 39 (1931) pp. 137–175.

17. William H. Bruvold, "Residential Water Conservation: Policy Lessons from the California Drought," in *Public Affairs Report Bulletin* vol. 19 (December 1978) (Berkeley, Institute of Governmental Studies, University of California); and Richard A. Berk, C. J. LaCivita, Katherine Sredl, and Thomas F. Cooley, *Water Shortage: Lessons in Conservation from the Great California Drought, 1976–77* (Cambridge, Mass., Abt Books, 1981).

18. Aaron Wildavsky, *Speaking Truth to Power* (Boston, Little, Brown, 1979), p. 134.

19. Giandomenico Majone, "Technical Assessment and Policy Analysis," *Policy Sciences* vol. 9 (1977) pp. 173–175.

20. Wildavsky, *Speaking Truth*, p. 136.

21. The historical statistics were revised by the Tucson Water Department in the spring of 1982 based on new population census data. The drop in use was previously thought to be from 204.6 gallons per capita to 139.5 gallons per capita. The revised estimate for the fiscal year 1978/79 low is 147.5 gallons per day. It is probable that the estimate of 204.6 gallons per capita in 1973/74, which was not adjusted, is also somewhat low.

2

Water Furor in Tucson

Physical Environment

The Tucson Basin is a broad alluvial valley in the Basin and Range province of southern Arizona (figure 2-1). In the northern part of the basin lies Tucson, a city of half a million people. Across the Tucson Mountains is the Avra Valley, an extensive agricultural area. The basin is drained by the Santa Cruz River, which runs northward through Tucson, leaving the basin at the Rillito Narrows, to the northwest of the basin. Just to the south of Tucson is the Papagos' San Xavier Indian Reservation. On the reservation, a cooperative owned by some of the Indians farms some 800 acres. Further south is a 5,000-acre pecan orchard lying along the Santa Cruz River near Sahuarita. To the west of this orchard, on the eastern and northeastern foothills of the Sierrita Mountains, are four copper mines where low-grade ores of copper are mined and concentrated, a process using large amounts of water. In the extreme south, on the Mexico border, is the small town of Nogales with a population of 12,000. The rest of the area is essentially sparsely populated desert. In 1980, the Arizona Groundwater Management Act created the Tucson Active Management Area (AMA) in order to begin coordination of the use and management of water among these various users. The Tucson AMA essentially comprises that area lying between the Santa Catalina and Santa Rita mountains on the east, and the Baboquivari and Silver Bell mountains on the west.

Estimated consumptive water use in the Tucson AMA as of 1980 is shown in table 2-1. Municipal and industrial water use was less than one-third of total consumptive use, but the share of those two sectors had been and still is rising steadily.[1] An increasing share of Tucson's municipal water demand has been met by retiring farms in the Avra Valley and transporting water over the Tucson Mountains. The displacement of agriculture is expected to accelerate in the future, and it is estimated by the Arizona Department of Water Resources that irrigated acreage will have declined by a third by the year 2025. During this same period of time, the Tucson population is projected to more than double, and may triple.[2] Consequently, municipal and industrial uses are expected in time to surpass agricultural water use.

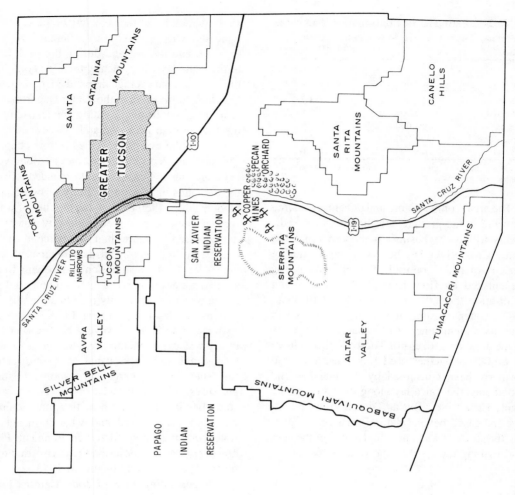

Figure 2-1. Map of the Tucson Basin, the Avra Valley, and the Altar Valley.

Table 2-1. Estimates of Consumptive Water Use Patterns, Tucson Active Management Area, 1980

Sector	Acre-feet	Percent of total consumption
Municipal	60,000	19.2
Industrial	30,000	9.6
Mining	53,000	16.9
Agriculture	170,000	54.3
Total	313,000	100.0

Source: Arizona Department of Water Resources, Tucson Active Management Area staff.

The area's only reliable and developed source of water is the groundwater borne by the sediments filling the basins; in modern times the Santa Cruz River has been dry except during periods of heavy rainfall. The groundwater is replenished by flow from the mountains and infiltration through the river beds. Until about 1940, water consumption in the area was less than its replenishment, or recharge. The farms in the area and citizens of Tucson—then a town of 60,000—together used less water than had formerly been consumed by the marshes and woodlands that once lay along the banks of the Santa Cruz River. But as Tucson grew, pumpage increased beyond natural recharge, depleting the stock of water accumulated over the ages and making the level of the water table fall.

Tucson Water Utility

The City of Tucson Water Department is the major water utility in the area, serving about 445,000 people both within and outside Tucson's city limits. A number of independent water companies provide pockets of separate service, although their numbers have declined in recent years as they have been bought out by the city utility. Since Tucson is growing rapidly, the city Water Department must continually enlarge its water system to meet the increased demand for water. Consequently, the city has a continuing need for revenues to finance improvements to its water system. Much of the capacity of the city's water system is needed to meet the high peak demand for water during the summer. Until

1976, demand over the peak day of the year was typically twice as high as the annual average demand, and the demand during the peak hour of the peak day was typically three and one-half times the annual average demand.[3] Consequently, a large proportion of the Water Department's costs resulted from providing the capacity to meet these high peak demands.

Prior to the events described in this chapter, Tucson managed its water resources along very traditional lines. Supplies were expanded to meet the demands of Tucson's rapidly growing population. From 1950 to 1970 the number of water service connections served by the city water utility increased by more than 400 percent. Water rates were kept low, and they decreased with higher levels of consumption. Tucsonans freely applied water to their arid environment, and per capita water use gradually rose to a high of 205 gallons per day in fiscal year 1973/74.

In the mid-1970s, however, the Tucson water utility was experiencing pressures on its ability to continue service as usual. A new direction for water policy emerged in response to these pressures. The importance of the Tucson case study lies in explaining how this innovation in water policy occurred and what it meant for water conservation in a desert environment. First, however, we must describe the events that took place. The case study begins in 1973.

The consulting firm of John Carollo Engineers was routinely hired by the City of Tucson to conduct capital-improvement and rate studies for the water utility. Until 1973 rates had remained very low and of standard form. However, in their report released in 1973, the consultants recommended substantial rate increases averaging 30 percent (40 percent in the mean use range for single-family residences) along with a mildly increasing block rate structure. (A block rate structure charges a different price for water in each additional block of use.) Previously, the block rate structure had been a declining one. Those recommendations were instituted by the city in February 1974. Despite the large rate increases, there was little outcry from the citizens, and per capita water use was not immediately affected. In fact, in the summer following the institution of the new rates, the utility reported the highest levels of water consumption ever for the city.

During the spring and summer of 1974, Tucson experienced one of the hottest, driest periods in its recorded history. On several days in June, the city Water Department's pumping capacity was unable to meet the afternoon peak demand. Some parts of the city, particularly the areas at highest elevations, ran out of water, and had there been a major fire, flow requirements for fire fighting could not have been met.[4] In July 1974 John Carollo Engineers was again retained by the Water Department to study the department's problems and to make recommendations for the improvement of the city's water system. The Carollo report, submitted to the Water Department in January 1976, recommended a six-year program of improvements to the water system in order to meet the continually increasing demand for water in Tucson—both for total quantities and for peak use.

Also during this period, an effort to set up a joint city-county management agency for water and sewer operations was underway. The Metropolitan Utilities Management Agency (MUM) was established in 1974 through cooperative agreements between the City of Tucson and Pima County. An advisory MUM Policy Board with members from the City Council, the County Board of Supervisors, and the general public was created. However, the City Council retained responsibility for the water utility budget and financing.

Political Environment

A group of political activists and environmentalists became a powerful faction of the Democratic party in Tucson in the early 1970s; its members were proving themselves to be formidable campaigners. Beginning with Ron Asta's successful bid for county supervisor in 1972, the "New Democrats" held ten other city and state elective offices by 1976. Two Astacrats, as they came to be called, were elected to the Tucson City Council in 1973. In the post-Watergate 1974 elections, six Astacrats waged successful campaigns for seats in the state legislature.

Finally, two more New Democrats were elected to the City Council in 1975.[5] As a result, by 1976 New Democratic council members Barbara Weymann, Robert Cauthorn, Doug Kennedy, and Margot Garcia had gained majority control over the seven-member Tucson City Council. Other council members were lone Republican Mayor Lewis Murphy and old-line Democrats Ruben Romero and Rudy Castro.

The new majority on the City Council and their New Democratic colleagues and supporters had run issue-oriented campaigns. With some minor differences among the candidates, their platforms were based on the ethic of controlled growth, honesty and openness in government, better mass transit, comprehensive planning, and the protection of Tucson's environment.[6] During their campaigns, they emphasized the need to curtail urban sprawl and revitalize the inner city. They also opposed all new freeway construction in the Tucson area.

Considerable controversy surrounded these liberal, environmentally oriented politicians. From the time Ron Asta first brought up the issue of controlled growth in his 1972 campaign, opposition to controlled growth became the rallying cry of much of Tucson's business community. Bumper stickers proclaiming "Construction feeds my family" and "No growth means no jobs" appeared with increasing frequency around town. In February of 1975, a group of developers, builders, and businessmen formed the Good Government League to provide political and financial support for the "pro-growth" point of view.

Controlled growth was a major issue in the 1975 council elections. Republican Mayor Lewis Murphy, in his bid for reelection, called efforts to develop a comprehensive plan for the city's growth "socialistic."[7] Materials distributed by the Good Government League extolled the virtues of a rapidly growing economy and an expanding job market. The New Democrats were accused of frightening away new industry and jobs and threatening people's right to choose how and where they wanted to live. The New Democrats countered with arguments about the costs of urban sprawl and unplanned growth.

The concerns of the New Democrats reflected their affiliation with the environmental movement. Issues such as the construction of freeways and urban sprawl reflected their more basic concern with protecting and preserving the Tucson environment. Water resources emerged as an improtant issue reflecting the fundamental environmental ethic of these new politicians. Water conservation represented a basic moral requisite that water not only should not be wasted, but also that it should be preserved to the extent possible. Moreover, some environmentalists in Tucson went beyond support of water conservation as a principle to the recognition of its utility as a planning tool. Some local opponents of the Central Arizona Project (CAP) argued that conservation would provide much of the additional water needed by Tucson in the future, without construction of the project.[8] The same people also argued, in a different context, that water resources would be an important limiting factor in future urban growth in Tucson.[9] The inconsistency of these two arguments is less important than the significance of the connection made between water resources management and political goals.

The hard-fought 1975 City Council election victories for the two New Democratic candidates were viewed as a mandate. County Supervisor Ron Asta, on the day following the narrow victories by his New Democratic colleagues, told the newspapers that "we have made the transition from challengers to leaders. We now have to produce. The Mayor is largely irrelevant."[10] As is summarized in table 2-2, the stage was set for a new scenario in water politics to be played out in Tucson over the next few months.

Development of Policy: January to June 1976

The task of translating campaign promises into substantive action loomed before the new majority on the City Council. They had campaigned for planning for controlled growth of the city and now were ready to act. They set about resolving the immediate pay raise and

Table 2-2. Key Events in the Tucson Scenario: Background Developments, 1972–1975

Year	Event
1972	New Democrat Ron Asta elected to County Board of Supervisors. Debate over controlled growth begins.
1973	Two New Democrats elected to City Council. Carollo firm hired to do study of water utility. Water rates raised substantially with little public outcry.
1974	Record high water use during summer exceeds water utility's well capacity.
	Carollo firm again hired to study water utility.
1975	Two more New Democrats elected, gaining majority control over City Council.

Table 2-3. Key Events in the Tucson Scenario: Development of Policy, January to June 1976

Month	Event
January	New Democrats take office on City Council.
February	Carollo report released calling for massive program of capital improvements, a 42 percent rate hike, and a cost-of-service approach to rates and charges.
April	New Democrats become deeply involved in Water Department rate and revenue deliberations.
	Council majority rejects Ron Asta's call for phased implementation of rate hike.
May	Deadline for Water Department budget nears. Pace of deliberations speeds up.
	Public hearing held with low participation, but clear opposition to proposed rate hike.
	Mayor Murphy comes out strongly against proposed changes.
June	New water rates adopted and instituted. Weather very hot and dry.

budgetary questions before the council so that they could address their own priorities of comprehensive planning and mass transit. Among the budgets to be considered was that of the Tucson Water Department about which the four New Democrats were to learn a lot more in the coming months.

The key events in the first six months of the New Democrats' tenure on the City Council are summarized in table 2-3. Water policy quickly assumed a place of unusual prominence in council deliberations. Although their perspectives and motivations varied, the New Democrats, the Water Department Staff, and the Carollo consultants joined forces to create a new direction for water management in Tucson.

The groundwork for collaboration and innovation in water policy between the utility staff and the New Democrats predated the latter's ascension to majority control of the City Council. Two council members, Margot Garcia and Robert Cauthorn, had served on the MUM Policy Board with fellow New Democrats Barbara Tellman and Ron Asta. The director of the Water Department, Frank Brooks, was closely aligned with the new politics faction of the Democratic party and had, in fact, been asked by them to run for mayor in 1975.[11] Although he declined to run for office himself, his ties to the new council majority were established before its tenure had even begun. In the new council, the staff of the Water Department had a receptive audience for their problems and the solutions they suggested. Although the consultant firm of John Carollo Engineers had been hired by the previous City Council, the consultants worked closely with the New Democrats and Water Department staff on the report that they were to present after the new City Council took office.

Water Department Staff Recommendations

The Carollo report, released in early February 1976, was immediately addressed by a joint meeting of the City Council, County Board of Supervisors, and Metropolitan Utilities Management Agency. As a prelude to discussing the recommendations of the Carollo report, the Water Department staff gave lengthy presentations on the problems faced by the utility. Staff presentations argued that rapid expansion of the distribution system in order to service the growing population was increasing costs more rapidly than it was increasing revenues. Serious capital-improvement problems were forecast by 1981. The city's chief hydrologist gave an overview of the complex well field and distribution system through which the city water supply was obtained. He pointed to rapidly declining water tables and the potential dangers of subsidence from central city wells as evidence that system expansion would have to be escalated.

Following these presentations the recommendations of the Carollo report were outlined. A massive six-year program of capital improvements to cope with the growing inadequacies of the utility's delivery capability was advocated. Financing for this construction was to come from a combination of bond sales, new and higher water rates, and system development charges. Although proposals for increases in delivery capacity were not unusual for a growing city like Tucson, the scale of these recommended improvements was out of the ordinary. An even greater departure from past practice came in the structure of Carollo's recommendations with regard to water rates. Not only were the proposed rates an average of 42 percent higher, they were predicated on the basis of cost of service, where operating and capital-improvements costs would be recovered from those customers responsible for costs.[12]

In supporting the rate structure recommendations of the Carollo report, the staff of the Tucson Water Department was thinking in terms different from the usual public utility preoccupation with simply providing an adequate and low-cost supply to all users. A staff memo to the City Council dated March 24, 1976, stressed sound water management and equitable financing.[13]

In the memo, the Water Department indicated its intent to "effectively manage the water resources of the Tucson and Avra Valley Basins." The city Water Department's ability to manage the goundwater of the Tucson and Avra valleys is limited because it has no control over pumping by other water users. Nevertheless, the Water Department staff recommended that additional sources of supply should be developed in Avra Valley, that pumpage by others should be reduced by the purchase of agricultural lands and selected private water companies, and that the use of wastewater for irrigation and mineral processing should be encouraged. Further, the department staff stated its objective was to "efficiently and effectively design and operate the water system."[14] Improvement of the distribution system and construction of three service reservoirs were recommended as the means to this end.

The staff intended to achieve more equitable financing through the adoption of Carollo's recommendations of setting rates on the basis of the costs of providing service. Customers at higher elevations were to pay lift charges for the costs of pumping to elevations higher than the city's well fields.[15] A system development charge was designed so that newcomers rather than the utility's existing customers would bear the cost of system expansion.[16] Increasing block rates were intended to place peaking costs upon large users.

The objectives of the Water Department staff and the New Democrats were matched on the matters of management and financing. However, there are reasons to believe that, for the benefit of the New Democrats, the staff may have added to its March 24 memo some objectives that the staff itself may not have shared.

The March 24, 1976, Water Department staff memo stated:

> The proposed water-pricing technique should be designed to discourage the wasteful use of the resource and seek to effectively manage the system demand during the peak period experienced through the hot summer months. An increased rate per ccf [hundred cubic feet] as increasing volumes of water are used should be applied to all customer classes according to the costs of providing water service.
>
> The proposed water rate structure and system development charge should discourage peripheral growth and encourage the development of vacant land within the metropolitan areas where supporting public services already exist or are much cheaper to provide than in the fringe areas.[17]

The new Democrats strongly supported this language because they felt that the new rates would encourage orderly development, eliminate water waste, and encourage conservation.[18] However, while conservation was a stated goal of the March 24 memo, the staff exhibited some confusion about conservation response to the higher rates.[19]

Although conservation was included as a stated "symbolic" tool of the new rates, it was not taken into account by the staff in revenue projections. Demand for water was considered completely inelastic; an increase in price would not affect the quantity of water purchased.[20]

At the same time, the staff continued to support, in theory, the potential for demand management through pricing strategies. The same memo in which the staff predicted unabated consumption with rate increases also suggested that the establishment of a surcharge on use during peak consumption periods might be another possible way of encouraging conservation.[21] This alternative was not pursued further in this or later staff work, but it did indicate support for demand management, at least in principle. It also emphasized that pricing was the only method of encouraging water conservation and curtailing peak demand that had been considered by the Water Department. No mention is made in either the memo or the accompanying report of the possibility of using educational campaigns or physical or legal restrictions on water use to achieve the Water Department's objectives.

The staff recognized that the overall 30 percent increase in revenue projected under the proposed rate structure would affect the various classes of customers quite differently.[22] The council was provided with tables presenting a cross section of dollar and percent rate increases for various classes of water customers at varying levels of water consumption and lift zones. The staff indicated that the change in monthly water bills under the new system would range from a decrease of 20 percent to a maximum increase of 185 percent for those customers using a lot of water and living in the highest lift zone. Other tables showed that very few customers would be affected by these highest lift rates.[23]

Portents of Trouble

While a certain amount of publicity accompanied the City Council's deliberations on the Water Department budget and rate structure, it was far from a topic of major public concern. A front-page headline in a local newspaper in January 1976 warned, "Plan to Raise Water Bill 400% Proposed by MUM Consultants."[24] This warning was shown to be grossly exaggerated in the accompanying text, however, and a follow-up story reported, "Water Rate Increase Now Put at 30 Pct."[25] Little else on the specifics of the proposed rate structure was reported in the local press prior to passage of the new rates in June. Local papers provided some coverage of the problems of the Water Department and some discussion of conservation ideas, although neither was treated as particularly important news.

In the meantime, the council majority continued to delve into the intricacies of operating the water utility. They became more and more convinced of the need to take fast, positive action to increase revenue and begin the necessary capital improvements. By March, however, County Supervisor Ron Asta was becoming equally convinced that the council majority was trying to do too much too quickly. On several occasions he asked Water Department staff to supply him with alternative lower budget and revenue scenarios.[26] Understandably reluctant to undermine a proposed budget favorable to their interests, the staff was slow in providing him with that information. Asta resolved to do his own computations, and on April 7, 1976, he presented the City Council with a memorandum suggesting substantial budget cuts for the Water Department. Asta, the leading figure in the New Democrat movement, had become nervous about the potential danger of the water issue. Although he personally agreed with the aims of the council majority, he had sensed the political danger of raising water rates so high so quickly. His suggestions met with a cold reception by the other New Democrats. MUM Policy Board member Barbara Tellman wrote Asta a scathing letter characterizing his memo as a "meat cleaver approach to cutting the budget."[27] Since County Supervisor Asta had no formal authority over city water rates beyond his advisory capacity as MUM Policy Board chairman, there was little else he could do. His suggestions to phase in implementation of the new rates or delay them until October also met with disapproval by the council majority.

From the perspective of the council majority, Asta's actions were particularly irritating. They had studied the problems of the Water Department long and hard, to the point where they were quite expert about the historical facts of its operation. They thought they understood the needs of the utility and were trying to come up with a sound budget and equitable rate structure based on the long-term needs of the community. They did not expect the rate increase to be popular, but they believed that the public would eventually accept the changes. In their view, Asta had not done his homework about the details of the utility, and was undercutting their efforts with his ill-founded and somewhat arbitrary proposed slashes in the Water Department budget.[28] The cooperative relationship between city and county officials in Water Department affairs had eroded, and the council majority forged ahead to achieve the new budget and revenue plan on schedule.

The exchange between Asta and the council majority indicates the low priority given to the problem of political acceptability by the council's New Democrats. They were willing to rely on a couple of public hearings to forewarn the public about the changes in water policy. At the same time, they understood that the public eventually would have to be educated about the water issue, so they included $50,000 in the water utility budget for public education. However, under the time constraints of the schedule for adoption of the new water rates, they put off further investigation into the details of a public education program until after the new rates had gone into effect.

Actions of the City Council

The schedule for review and adoption of the water utility budget and financing had been established by the mayor and council in January of 1976. It called for a public hearing on the proposed changes on May 11 and final adoption of the budget and rate structure on May 24. Since the final details of the dramatic changes proposed for financing the water utility became available only toward the end of April, the pace of deliberation on the proposed budget and rate structure became quite hectic in May.

A public hearing on the proposed water utility budget and new rate structure was held May 11, 1976. The primary concern of most of the people speaking was the proposed rate increase. Little or no understanding of the different elements of the rate structure or the proposed uses for the increased revenues was evidenced by the speakers representing the public. What did come across from their remarks was a disbelief that any water problem, much less an impending crisis, existed in Tucson.

By the May 24, 1976, meeting of the council, many questions and problems had not yet been resolved. At that meeting Vice Mayor Weymann explained that the rate decision was being postponed so that changes could be made to help single-family residents who were going to bear the greatest burden under the new rate structure. She said that the rates had already been reduced four times from their suggested levels of a 42 percent increase down to only a 22 percent increase and that the city had come to the point where water service would be threatened if further reductions were made. She concluded by saying that the council members realized that what they were doing was "very, very unpopular," but that they were doing "what they know has to be done for the community."[29]

During the ensuing discussion, each member of the council and the mayor commented at length on the proposed changes. Council member Garcia remarked that she found it amusing that the people who usually most vociferously defend the free enterprise system were asking that water use be tied to taxation rather than to cost of service, which allows each individual to use as much water as he or she is willing to pay for. Council member Kennedy commented that the council was continually being told to run the water utility like a business, but then was also told that costs should be passed on to everyone equally instead of being charged to whom the costs were attributable as would be done by any business. He went on to say that he was particularly disheartened by the lack of public understanding of the problems faced by the county and the city in water policy.

Mayor Murphy countered that the people of the community had made clear their opposition to the budget and proposed rate structure. He went on to say that although the capital improvements proposed in the budget might be needed, the changes should not be made so quickly in an effort to catch up with past failures, especially when the people using the system were not satisfied with the changes. Council member Castro explained that he would vote against the budget and rate increases because most of the people he represented did not have jobs—"How are they to pay these rates?"[30] Council member Kennedy stated that he was very disappointed because Murphy had not made one constructive suggestion as to how to change the budget during all of their months of deliberations. Mayor Murphy replied that "sometimes it is necessary to do what your constituents want you to do."[31] Further discussion on the rate changes was postponed until the next regular meeting, at which time a revised rate structure was to be voted upon.

Later during the same May 24th meeting, which turned into a 13-hour marathon, the water issue came up again. In a move separate from the rate-making controversy, council majority member Cauthorn proposed an ordinance that would give the city manager power to declare "water alerts" and which would call on Tucson residents to stop outdoor watering voluntarily on Wednesdays in June and July. The water alerts would empower the city manager to restrict outdoor watering for as many as two days a week in order to maintain adequate water pressure and fire-fighting flows. A fine of $2.50 for each violation was included in the proposed ordinance. The Cauthorn proposal was adopted by the council majority after a brief but heated discussion in which Mayor Murphy charged that the ordinance had been "dropped out of the clear blue sky without an opportunity for community digestion or reasonable debate."[32]

The "Waterless Wednesdays" program was labeled a "peak season restriction." It called on residents not to "fill or partially fill a swimming pool or water outside plants on Wednesday during the months of June and July."[33] Compliance was left on a voluntary basis in 1976, implying that in the future it might become mandatory. Neither this voluntary program nor the water alert restrictions were presented as conservation measures—they were measures to reduce the load on the delivery system. They were, however, efforts at demand management, and as such were a significant break from past practices.

At the June 7, 1976, meeting of the Tucson City Council, the final revised water rates were presented by city staff. City Manager Valdez explained that the revised rate structure represented a reduction in revenue from the single-family residential class and an increase in revenue from the commercial and industrial customer classes. It was shown that under the revised rate structure "customers will experience anywhere from a 41 percent decrease in charges to an 85 percent increase in charges *if they use 6 ccf* [hundred cubic feet] *per month*" (emphasis added).[34] It was not explicitly mentioned that very few people used as little as 600 ft^3 (748 gallons = 100 ft^3) per month in the summer. As is indicated in chapter 5, about 2,000 ft^3 was the summer average.

The council majority was aware that the proposed rates would affect customers in different locations and with varying patterns of water consumption quite differently. After all, that was the idea. They had no idea, however, that the level of impact when the rates were implemented would be as extreme as they were.[35] But Mayor Murphy, at the June 7 meeting, again expressed opposition to the new rate structure, calling the burden on single-family residences excessive and warning that "the high water user should, indeed, be charged for the amount of water he uses, but, in my judgement, he should not be overcharged on the theory that the more you charge incrementally, the less the consumer will use."[36]

The council majority members reiterated their positions that the new rates were in fact more equitable than any previous ones had been, and that issues more important than public popularity were involved:

It is important that the progressive rate structure that is reflected in this ordinance comes much closer to a fair allocation of costs in the system than any rate structure that the City has previously had. In addition to allocating the costs more fairly than past rate structures, it also contains in it a premium, a recognition that the excess and unmetered or unwarranted use of water in the community simply is too costly to be continued. There must be some kind of incentive in the rate structure for people to minimize

the use of water, especially in the case of outdoor watering.[37]

Those people who want to have a green jungle around their house even though they live in the desert, will have to bear that economic cost.[38]

Thus, with a touch of righteousness and a lot of naiveté, the four majority members of the Tucson City Council adopted the new water rates over the objections of the three other council members. The new rates were to go into effect in July, the following month.

Thus the newcomers to the Tucson City Council adopted a water management policy that would have more than ordinary repercussions. Although they instituted rates and charges based on notions of demand management, they still planned for a delivery system to meet any and all demands. Revenues raised were to be earmarked largely for land purchases in Avra Valley to gain the right to use the water and to finance a delivery system from the source to the Tucson area. This was deemed necessary to meet the level of peak usage in the summer due essentially to outside watering demands. In fact, only slightly above 30 percent of the water delivery system in 1976 was required to meet the average daily water demand, with the remainder serving peak daily and peak hourly demands.[39] As one of the newcomers to the council argued,

This (outdoor watering) is the kind of tremendous extra use in the summertime which places such a burden on the water system, and which demands the rapid increase in capital in order to build a system to *deliver as much water as anyone could possibly want at any time* [emphasis added].[40]

If they had not assumed the necessity of meeting peak demands or had adopted a nonpricing policy aimed specifically at reducing peak demands, the cost to any particular customer might have been more acceptable. In retrospect, two of the newcomers recognized this problem: "The problem with the Water Department that had a spill-over effect on us was that they wanted to build the perfect gold-plated system—one large enough and perfect enough to meet all possible contingencies,"[41] and "The problem with the people in the department was they were only concerned with satisfying peak demand."[42]

At the time, however, the council majority had little idea of the profound implications of their actions.

Electoral Response:
July 1976 to January 1977

The new power of the city manager to declare water alerts made front-page news the day following its adoption. In fact a water alert was never declared, and that potential water use restriction quickly faded into oblivion. "Waterless Wednesday" continued through June and July with minimal fanfare and was declared a success by the water utility director.[43] The actual effectiveness of the Waterless Wednesday was difficult to assess, since it only occurred six times and other factors intervened—on the first Wednesday (June 23) it was cloudy; on the second Wednesday it rained; and by the third Wednesday, the impact of the new water rates had been felt, and citizens were in no mood to cooperate with city government.

Reaction to the new water rates used in the water bills mailed out in July 1976 was immediate and intense. For a public plagued by rising taxes in an inflationary economy, a significant increase in the cost of water was too much to tolerate. A local television store owner, at his own expense, took out a full-page advertisement in the local newpaper, calling for contributions to a "Citizens' Water War Chest." This group descended on the City Council and forcefully demanded a public hearing on the new water rate, a "roll-back" of the new rate structure, a complete housecleaning in the Water Department, and, most importantly, the resignation of those council members who had voted for the new rate structure. These events are summarized in table 2-4.

The dramatic increases in some people's water bills caught the council majority members by surprise. They had expected moderate increases in charges to most customers and perhaps a few instances of quite drastic increases to profligate water users.[44] Instead, although most Tucson-

Table 2-4. Key Events in the Tucson Scenario: Electoral Response, July 1976 to January 1977

Month	Event
July	First water bills under new rates are much higher than anticipated.
	Public outcry ensues, and recall drive against New Democrats is begun.
August	New Democrats backtrack—rescind lift charges and establish lifeline rate.
September	Recall petitions are filed bearing twice the necessary signatures.
October	Slate of candidates promising to roll back water rates is endorsed by recall organizers.
	New Democrats hold water workshops, but attendance is poor.
November	Citizens' Water Advisory Committee (CWAC) is appointed by city manager.
	Focus of recall campaign broadens beyond water issue.

ans did only receive a moderately higher water bill, a substantial number of people, particularly in the more affluent parts of Tucson, received a bill that was drastically higher. Even the Water Department staff, which had prepared numerous statistical tables to show the projected impact of the new rates on utility customers, was not prepared for the extremes encountered in the July billing.[45] No one had anticipated the high levels of consumption by Tucsonans during the unusually hot and dry month of June. The bills of some people in high lift zones actually quadrupled from June to July, while the bills of many others doubled.[46] Those who did not personally experience a dramatic rise in their water bills were in any case aware of others who had.

The complex and confusing rate structure devised by city staff and the council majority in their efforts to be as fair and equitable as possible made it almost impossible for water customers to recompute their own bills. This structure tended to reinforce the widely held perception that the council's actions were capricious and unnecessary.[47] Within two weeks after the new bills were mailed out in July, a drive to recall the council majority members had begun.

In an effort to quell the violent opposition engendered by the higher rates, the council majority announced early in August that the new lift charges would be rescinded. They also aided the low-income, low water user through a life-line rate of $3 for all customers using 400 ft^3 or less per month.[48] These concessions had little effect on the recall drive however, particularly when four days later the council majority rejected a proposal by the other council members that the entire rate hike be rescinded. By the middle of September, twice the number of signatures necessary for a recall election had been filed.

Throughout the summer and fall of 1976, water remained the major local issue in Tucson. The morning paper, the *Arizona Daily Star*, generally supported the council majority in its editorials, calling the "Rate Structure Sound but Charges Too High."[49] It was argued in several editorials that water itself should be valued and that the higher rates would encourage conservation.[50] Despite this editorial support, the recall drive and its leaders, John Varga and John Fitzgerald, dominated the news coverage of the water issue. The *Tucson Citizen*, the evening paper, came out against the council majority, and the controversy continued to rage. Both of the local papers ran series of background reports on the Tucson water situation.

Both sides of the recall drive engaged in hyperbole. MUM director Frank Brooks warned the Tucson Trade Bureau that without revenue from the lift charges, the city faced low water pressure, no new hookups in some areas, possible water rationing, and a loss to local businesses of $26 million.[51] John Fitzgerald in turn accused the council of "using the same scare tactics the oil companies did."[52] The recall leaders took out a full-page ad on the day of a council meeting urging the public to join them at city hall for a "showdown." The ad asked the four council Democrats to rescind the water rates and resign immediately.[53]

During the late summer and fall of 1976 a Citizens' Recall Committee was formed. The committee interviewed numerous individuals who had expressed a desire to run against the council members facing recall. The Citizens' Recall Committee eventually gave its endorsement to three persons who, if elected, pledged to roll back the water rate as their first order of business. Their fourth endorsement went to a can-didate who only promised to study the water situation fully. Other groups also became involved. One, the Good Government League, composed of developers and Chamber of Commerce members, endorsed three of the same candidates that the Citizens' Recall Committee endorsed.

The council members facing recall attempted to explain the necessity for their actions on the water rates. A number of water workshops were put on by the council majority members in an effort to get after-the-fact support for their new water policy. But attendance at these workshops was low, and much of the loss of public faith and support in the integrity of the new water policy and its authors seemed irreversible. The council members faced hostile constituents as they walked their wards, and reasoning with them was not always productive. For example, later in the fall when water use had declined with cooler weather and water bills had gone down accordingly, people would tell a campaigning council member that none of this would have happened if current lower rates had been in effect during the summer. When the council member would try to explain that the same rates were still in effect, the angry constituent would often refuse to believe it.[54]

In November, a Citizens' Water Advisory Committee (CWAC) was appointed by the city manager to look into the needs and operation of the Tucson water utility. The members of the committee were chosen to reflect a cross section of the community that had not previously been actively involved in the water controversy. They were instructed to pursue a nonpartisan approach to understanding the needs of the water utility and to make recommendations to the city government following the recall election. A new consulting firm was chosen to replace John Carollo Engineers. Black and Veatch, an internationally known firm, was selected to reexamine the recommendations of its predecessor.

The appointment of the CWAC task force marked the beginning of a gradual decline in the emotional debate over water. Although still an issue, water was increasingly minimized by the recall challengers. "The city's water rates are just one example of the Democratic City

Council majority's arrogant disregard of what people want," proclaimed one of the candidates.[55] By mid-January, shortly before the election, all recall candidates had backed down from promising that water would ever again be cheap, although three candidates on the "official" recall slate of four did continue to promise to roll back the water rates and study the problem.[56]

The council members facing recall hoped that this backpedaling by their opponents would cast a better light on their own positions. They also hoped that CWAC would tend to defuse the water issue politically. If public outcry had been dampened by the passage of time and these other factors, the four council majority members hoped to withstand the recall. They were to be rudely disappointed.

On January 8, 1977, the recall election was held. The three council members who faced recall (one of the four who had voted for the new rates resigned and accepted a job out of state) were soundly defeated. Each candidate endorsed by the Citizens' Recall Committee was victorious.

Public Acceptance of Higher Rates

Actions of the New Council

On the day following their landslide victory, the newly elected council members reiterated their commitment to roll back water rates.[57] (Post-election events are summarized in table 2-5.) At the first meeting of the new council, however, a lengthy memo from the city manager spelled out the dire implications of such a move. The memo concluded that "any further reduction in revenues at this time will have the effect of requiring greater increases in rates in the future."[58] This message was reinforced by a February 4, 1977, memo by Finance Director James Kay, Jr., on the effects of the proposed rollback. In it he recommended that if a rollback was to be instituted, it should be followed immediately by a 26 percent rate increase and another substantial increase within three months, or the

Table 2-5. Key Events in the Tucson Scenario: Following the Recall Election, January to June 1977

Month	Event
January	New Democrats are defeated easily in recall election.
	New City Council stalls on rate rollback after strong warnings from staff.
February	CWAC endorses increased water rates and conservation program to reduce peak usage.
	Public hearing brings support for increased rates from business groups.
March	City Council passes even higher water rates based on cost of service, but eliminates lift zone charges.
April	"Beat the Peak" public education campaign is devised with strong support from local media.
June	"Beat the Peak" campaign begins.

city's bonding capacity would be affected adversely.[59]

Members of the Citizens' Water Advisory Committee and the Water Department staff convincingly explained to the recall winners that higher rates were necessary to pay for the city's growth and expansion during the previous fifteen years. There was a great deal of "catching up" required. The new council members soon recognized that increased rates were necessary for improvements in the existing water system, and to finance the purchase of additional supplies in Avra Valley and a distribution system to deliver those supplies to the metropolitan area. Ironically, the recall winners discovered that the recalled new politics faction had actually raised rates to gain water supplies and build distribution systems to meet expected growth, not to limit growth as had been commonly assumed. One of the winners privately admitted that the new politics faction "did the right thing. The problem was they tried to do it all at once."[60] Publicly, council member Cross admitted that "I am certainly a lot wiser now than I was then (when I promised to roll back rates). . . ."[61] Her colleague James Hooten declared that "I have found that [an immediate rollback] would not be in the best interests of the city."[62] This backtracking by the newly elected council members led the morning paper to editorialize that in reaching conclusions the same as those of the previous council, the new officials were making a "tardy endorsement" of their predecessors' ideas.[63]

The new council members had stalled any decision to roll back rates prior to the release of the *Citizens' Water Advisory Committee Report* in early February.[64] When the report was presented to the mayor and council, it became the basis of most of the subsequent debate over what actions the council should take. The CWAC endorsed "cost of service" as a general principle to be followed in setting rates. Most of the committee's recommendations echoed the changes made by the previous council, including flat winter rates, increasing block summer rates, and lift charges. The rates proposed by the committee assumed a reduction in peak usage of 25 percent resulting from a conservation program. Even assuming this conservation response, the proposed rates represented an average 10 percent increase over the existing rates passed by the previous council, with suggested increases averaging 13 percent for the following year.

The need for a public education campaign to reduce summer peak demands was also urged in the report. The CWAC was careful, however, to point out "that any and all conservation measures it recommends are solely in the interest of controlling capital costs and customer charges. The committee specifically disavows any intent to sound an alarm as to the adequacy of water supply."[65] The primary purpose of such a program would be the conservation of capital rather than of the water resource itself. In other words, if peak summer demand could be reduced, then the capital improvements necessary to meet growing demand could be postponed. Consequently, the impact of necessary water rate increases could be delayed and lessened.

Attendance at a public hearing on the proposed water rates indicated that interest in the question of water rates was still very much alive.[66] More than 200 people attended a meeting on February 28, 1977, which dragged on for more than 4 hours. Most of the speakers representing the public were quite specific in their remarks, commenting on particular features of the proposed rate structure. But speakers were generally not as emotional as at public hearings on the volatile water rate issue over the past year.

The partisan political nature of much of the involvement of the business community in the recall election was exemplified by the testimony of the president of the Tucson Chamber of Commerce at the hearing. Although the Chamber of Commerce had actively opposed the New Democrats' efforts to institute these charges, the group's president spoke at length in favor of essentially the same rate structure as the one before the new City Council. He expressed strong support for the cost-of-service approach and the need to stress voluntary conservation to prevent greater rate increases in the future. He also expressed support for the three lift zones proposed by the CWAC, although the Chamber had been among the most vociferous opponents of the eight lift zones used by the previous City Council. The Chamber president ended his remarks by assuring the newly elected council members that "the chamber understands the dilemma this places on honorable people. It further believes that one of the many benefits of a democracy is the privilege of change, when facts dictate that it is in the best interests of the people."[67]

Following this generally low-key public hearing, the mayor called a special meeting of the council on March 3, 1977, to adopt a water rate ordinance. The newly elected council and their still wary colleagues refused to accept the recommendations of the CWAC for lift zones, and these were eliminated from the ordinance. No objections to the cost-of-service principle underlying the rate structure were expressed. Neither did any of the council members question the need for additional revenues, or the desirability of engaging in demand management strategies.

Council member Cross argued that for the revenue expected to be generated (about $100,000 annually), the lift zones were not worth the ill will they would cause.[68] Leaving the lift zones out gave the recall winners something to offer the people who had worked so hard to return Tucson water policy to the status quo. Instead of lift charges, an isolated-area surcharge was imposed for water service in areas not contiguous to the main urban system. Since these isolated areas were not within the city's corporate limits, residents of these areas had no vote in city elections. The mayor ended the discussion by asking his colleagues on the council "to put devisiveness behind them and . . . adopt the ordinance unanimously."[69] The ordinance passed by a vote of six to one. In explaining his "no" vote to the press after the meeting, council member Amlee explained, "The whole rate structure is based on the assumption that we are going to reduce water consumption by 25 percent. That's just a guess. The whole thing is too flimsy. What if we don't conserve 25 percent?" [70]

Fewer than six weeks after taking office, the recall winners had joined the other members on the City Council in raising water rates above those set by the recalled council majority. They had even committed themselves to three more raises in as many years. The leaders of the Citizens' Recall Committee and their supporters were livid. They felt betrayed and subsequently called for the winners' resignations. There was even talk of another recall election. Nothing came of such talk, however, and the new council was able to turn its attention to other matters.

Selling the Public: Beat the Peak

The capital-improvement projections upon which the new water rates were based assumed that peak outdoor water usage by Tucsonans would be reduced. It remained for the city staff and elected officials to develop, fund, and implement a successful voluntary demand-management program.

A $50,000 water education and media budget was designed, based upon the money already allocated by the previous City Council. Approximately one-fourth of the budget was to be expended in presenting a general background of the Tucson water situation to the public. The rest of the money was to be spent in explaining and advocating the "Beat the Peak" summer peak demand-management program. Under this program, during the summer months residents were to be urged never to water lawns and other outdoor plants more than every other day, not to water on Wednesdays, and never to water between 4:00 P.M. and 8:00 P.M. An often-

quoted, but unsubstantiated statistic used during the discussions on the media budget was that the expenditure of each dollar in advertising was projected to save $1,000 in future capital expenditures.[71] Despite this impressive "benefit-cost" ratio, the council members were concerned that the program was more expensive than necessary. It was suggested that the CWAC and the mayor and council meet with a group representing the local print and electronic media to see what could be accomplished on a public service basis. The staff was instructed to revise the budget downward and to arrange for such a meeting between media representatives and city officials.

On April 1, 1977, the entire City Council and three members of the CWAC met with representatives from the two local papers and every Tucson television and radio station, as well as the manager of a local advertising agency. Mayor Murphy began the meeting by describing the general outline of the message the city hoped to get across to the Tucson public. He solicited their cooperation in what he termed "the single most important educational undertaking that we have had, certainly within the last 15 years, if not in the history of the city."[72]

By the time the meeting ended, there was a commitment expressed by all the electronic media people present to support a public education campaign with donated air time and production services. The radio and television delegation expressed the concern since their time and effort would be donated, it would be unacceptable to have any tax money spent on newspaper advertising or billboards. An advertising executive at the meeting pointed out that billboards could also be donated, although production expenses might have to be paid for. The newspaper representatives responded that they could not guarantee free display advertising space without approval from their publishers, but they did offer front-page information boxes giving reminders about Beat the Peak and perhaps some statistics as to peak pumpage or reservoir levels. They also assured the city officials that news coverage of the water issues would continue at a high level. The spokesperson from the *Arizona Daily Star* in turn asked the city officials to be very careful in the information to be relayed to the media so that the campaign would remain accurate and credible. It was agreed that the city Community Relations Department would spearhead the campaign with the help of the advertising executive in attendance at the meeting. During the meeting, it was estimated that, with the donated advertising space, air time, and production services, the campaign would cost less than $5,000.

From this auspicious beginning, the Beat the Peak campaign in Tucson quickly developed into an annual multimedia endeavor lasting from June 1 to August 31. The campaign has relied primarily on television and radio spots, billboards, and water bill inserts. Other techniques that have been incorporated into different annual versions of the campaign include tent cards for restaurants, buttons on Water Department employees, television weather forecast reminders, fliers in grocery bags, and postage metering ads. The Beat the Peak speakers' bureau has also taken its demand-management message onto many public affairs television programs and before many groups around town.

Beat the Peak was designed and sold to the public as a way to defer capital expenditures by reducing peak demands on the system. Tucsonans were told that although their water rates would continue to increase, they would increase more slowly with public cooperation to reduce outdoor watering on summer afternoons. Beat the Peak was never tied to the idea that an individual could personally save money by using less water under an increasing block rate structure.

Results of Policy

The effectiveness of demand-management policy in Tucson is indicated by some significant changes in aggregate water use patterns over the last few years. Per capita pumpage changed from 189.4 gallons per day as an average during fiscal year 1976/77 to a low of 147.5 in 1978/79. The reduction in peak-day pumpage was even more impressive. The largest amount pumped per day in 1976/77 was 131.1 million gallons. The peak-day pumpage had been reduced to 113.2 million

gallons by 1978/79.[73] On the basis of this evidence the city's water program, particularly Beat the Peak, was widely heralded a success. Reduced use by single-family residences and the decline of outdoor watering were identified as the major contributors to these savings.[74] However, by 1980 there was evidence that both per capita use and peak-day use were climbing toward the levels established before the events described in this chapter took place. Further, some critics began to question whether what the city was doing was really water conservation.[75] The next chapter discusses concepts of conservation and the roles they played in Tucson.

Notes

1. Total pumpage was estimated at 395,000 acre-feet per year. Incidental recharge from agricultural, municipal, and mining use was 73,000 acre-feet. Including the use of 9,000 acre-feet of sewage effluent, consumptive use totaled 313,000 acre-feet.

2. Estimates were made by the Tucson Active Management Area staff of the Arizona Department of Water Resources.

3. See table 4-1 in chapter 4. Estimates are from John Carollo Engineers, "Report Submitted to the Metropolitan Utilities Management Agency," Final report (Tucson, Ariz., March 11, 1976).

4. As our analysis later shows, the effects of the price increase may have simply been masked by effects of the record heat wave. Per capita use fell in the following fiscal year, although not to the pre-1973/74 level.

5. Barbara Weymann and Robert Cauthorn were elected to the City Council in 1973; Lucy Davidson, Sue Dye, Morris Farr, Jo Cauthorn, Sister Clare Dunn, and Bruce Wheeler were elected to the state legislature in 1974; and Doug Kennedy and Margot Garcia joined the City Council in 1975.

6. Ben MacNitt, "Democrats Sweep Council Races," *Arizona Daily Star*, November 5, 1975, p. A-1.

7. MacNitt, "Democrats Sweep."

8. Interview with Barbara Tellman, Metropolitan Utilities Management Agency Policy Board, Tucson, Ariz., April 17, 1980. A 1974 report written by City of Tucson Water Department staff in opposition to the CAP did not mention the issue of water conservation at all.

9. Interview with Ron Asta, Pima County Board of Supervisors Tucson, Ariz., April 28, 1980.

10. MacNitt, "Democrats Sweep."

11. Interview with Frank Brooks, Metropolitan Utilities Management Agency, Tucson, Ariz., November 8, 1980.

12. Tucson City Council, Pima County Board of Supervisors, and Tucson Metropolitan Utilities Management Agency, "Minutes of Joint Meeting" (Tucson, Ariz., February 9, 1976). In the ten years after 1966, the number of service connections had increased by 72 percent, production facilities had more than doubled, and the distribution system had gone from 1,300 miles of main to 2,500 miles, or an increase of 96 percent. Although revenues had increased during this time, the Water Department staff argued that costs had increased much more rapidly and that projected revenues would be insufficient to meet projected costs for fiscal years 1976/77 and 1977/78.

13. Tucson Water Department, "Memo to Mayor and Council" (Tucson, Ariz., March 24, 1976) p. 2.

14. Ibid., p. 4.

15. This cost was to be recovered by a surcharge of $0.20 per 100 cubic feet (ft^3) per lift zone of water delivered above the base service area (Lift Zone 0). Since eight lift zones were proposed, this arrangement meant that some customers would pay surcharges of $1.60 per 100 ft^3 of water delivered.

16. The purpose of the system development charge was to recover the *total* cost of increasing the capacity of the city's new water delivery system by levying a charge upon new customers. The Carollo report indicated that the cost of the proposed improvements would average $690,000 per million gallons per day of plant capacity. This unit capital cost was multiplied by the average peak-day demand per customer of 0.00105 million gallons per day to give a capital requirement per customer of $725. This amount was recommended by Water Department staff as the system development charge for new customers with ¾-inch connections (the average single-family dwelling). The system development charges for those with larger service connections were calculated using the relative capacities of the meters installed.

17. Tucson Water Department, March 24, 1976, memo, pp. 6–7.

18. Interview with Margot Garcia, Tucson City Council, Tucson, Ariz., April 14, 1980.

19. Interview with Frank Brooks, Metropolitan Utilities Management Agency, Tucson, Ariz., November 4, 1980.

20. Tucson Water Department, "Memo to Mayor and Council" (Tucson, Ariz., April 30, 1976) p. 5.

21. Ibid.

22. Ibid.

23. Ibid.

24. David Hatfield, "Plan to Raise Water Bills 400% Proposed by MUM Consultants," *Arizona Daily Star*, January 30, 1976, p. A-1.

25. See "Water Rate Increase Now Put at 30 Pct.," *Arizona Daily Star*, February 4, 1976, p. A-1.

26. Asta interview, April 28, 1980.

27. Tellman interview, April 17, 1980.

28. Garcia interview, April 14, 1980.

29. Tucson City Council, "Minutes" (Tucson, Ariz., May 24, 1976) p. 23.

30. Ibid., p. 52.

31. Ibid., p. 53.

32. Ibid., p. 75.

33. See *Tucson Code*, Ordinance No. 4490. Adopted May 24, 1976.

34. Office of the Mayor and Tucson City Council, "Mayor and Council Communication" (Tucson, Ariz., June 7, 1976) p. 2.

35. Garcia interview, April 14, 1980.

36. Tucson City Council, "Minutes" (Tucson, Ariz., June 7, 1976) p. 7.

37. Ibid., p. 11.

38. Ibid., p. 9.

39. See the City of Tucson's *Metropolitan Utilities Management Agency Report* (Tucson, Ariz., April 30, 1976).

40. Tucson City Council, "Minutes," June 7, 1976, p. 8.

41. Off-the-record interview, 1980.

42. Off-the-record interview, 1980.

43. See "Wednesdays Did Make A Difference," *Arizona Daily Star*, August 4, 1976, p. A-1.

44. Garcia interview, April 14, 1980.

45. Brooks interview, November 4, 1980.

46. See editorial, "Rate Structure Sound but Charges Too High," *Arizona Daily Star*, July 31, 1976, p. A-10.

47. Interview with Stephen Davis, Tucson Water Planning Division, Tucson Water Department, Tucson, Ariz., October 31, 1980.

48. Interview with Barbara Weymann, Tucson City Council, Tucson, Ariz. April 14, 1980.

49. "Rate Structure," *Arizona Daily Star*.

50. Water itself was not assigned a value. Only the costs of extraction from the ground and distribution to the consumer were included within this rate system.

51. Julie Tripp and Bob Lowe, "Water Cuts May Begin Next Year," *Arizona Daily Star*, August 20, 1976, p. A-1.

52. "Recall Leaders Reject Water Warnings," *Arizona Daily Star*, August 21, 1976, p. A-1.

53. Ibid.

54. Garcia interview, April 14, 1980.

55. "Incumbents Branded 'Arrogant'," *Arizona Daily Star*, December 29, 1976, p. B-11.

56. Ken Burton, "Challengers Avoiding Cheap-Water Promise," *Arizona Daily Star*, January 16, 1977, A-1.

57. Chuck St. Cyr, "Jubilant Election Winners Pledge Responsiveness," *Arizona Daily Star*, January 19, 1977, p. A-10.

58. Joel D. Valdez, "Memo to Mayor and City Council" (Tucson, Ariz., Office of the City Manager, January 28, 1977) p. 7.

59. James Kay, Jr., "Memo to Valdez" (Tucson, Ariz., Tucson Department of Finance, February 4, 1977).

60. Off-the-record interview, 1980.

61. Carol Stengel, "Council Members Put Off Action To Cut Water Rates," *Arizona Daily Star*, January 25, 1977, p. A-1.

62. Ibid.

63. "Tardy Endorsement" (editorial), *Arizona Daily Star*, January 29, 1977, p. A-10.

64. City of Tucson, *Citizens' Water Advisory Committee Report* (Tucson, Ariz., 1977).

65. Ibid., p. 5.

66. A memo from the mayor and council released prior to the hearing outlined a rate structure very much like that recommended in the CWAC report. It included three lift zones, a flat winter rate, and an increasing block rate in summer. It did not include a lifeline rate or system development charge, also along the lines of the CWAC report. Finally, the memo called for "a summer water demand management program of voluntary alternate day watering to achieve a 25 percent reduction in outdoor watering during the summer peak period." This basic program was to be accompanied by a more general, comprehensive, continuing public education program on water matters.

67. Tucson City Council, "Minutes" (Tucson, Ariz., February 28, 1977) p. 15.

68. Tucson City Council, "Minutes" (Tucson, Ariz., March 3, 1977).

69. Ibid., p. 3.

70. Chuck St. Cyr, "Council Oks Water Rates," *Arizona Daily Star*, March 4, 1977, p. A-1. Rather than a 25 percent reduction in consumption, the program was intended to achieve a 25 percent reduction in outdoor watering during the summer peak period.

71. Tucson City Council, "Minutes" (Tucson, Ariz., March 28, 1977).

72. Tucson City Council, "Minutes of Water Education Meeting" (Tucson, Ariz., April 1, 1977) p. 4.

73. R. Bruce Johnson, "Changes in Water Use—Tucson, Arizona." Paper presented to the American Water Works Association, Phoenix, Ariz., 1979.

74. Davis interview, October 31, 1980.

75. Charles Bowden, "Beat the Peak Quackery," *Arizona Daily Star*, August 2, 1981, p. C-3.

3

Preachments in the Name of Conservation

There is more sermonizing about water than about any other natural resource. The appeal of water to humans is far more spiritual and emotional than intellectual. Aesthetic sensibilities are closely bound to the presence of water. In nature, a shimmering brook, a majestic waterfall, or a placid pond are beautiful. Since Greek and Roman times decorative water fountains have refined urban settings. Water also is tied closely to righteousness. Baptism is common in many religions. The allocation of water is commonly an ethical choice, and in many cultures the power to divide water is given only to priests or judges who are trusted to serve equity.[1] Access to water is regarded as a moral right, and discriminating among claimants to water on the basis of wealth or position is in many places regarded as immoral. Humans associate availablity of water with survival at a nearly elemental level of feeling. Drought, along with famine and pestilence, is an age-old threat against which it is inconceivable to have too much security. Risk aversion to water shortage is extremely high, and even the smallest chance of doing without or with less is objectionable.

Because of these deeply held feelings about water, modification of water allocation policy must be supported by viable preachment. Symbolic appeals often become more important in water policy than rationally based analysis. Ideologies, or abstract values and preferences to which symbols allude, serve as political organizers. In the name of ideology, policymakers may be motivated to take action, the political risks of which might not be otherwise justified. Commonly held ideology can serve as the basis of political coalition-building necessary to deliver the majorities required in a lawmaking process. Further, ideology can serve a crucial role in policy implementation by providing the rationale for demanding sacrifices from the target groups and individuals who could not be induced to cooperate without an appeal to communal values.

Not all ideologically based appeals positively serve the needs of policymaking and implementation. It is possible for policies based on certain moral ideals and emotional preferences to encounter serious problems. Symbols may mean different things to different groups, and conflicts may arise over the true meaning. Preachments may be so different from perceptions of reality that they are not credible to policymakers or to the public. Lack of credibility occurs when the operational aspects of policy are so divorced from ethical aims that values do not appear to be served. Further, preachments may be at odds with other policy imperatives. Ideology can result in politically dysfunctional behavior in policymaking when it dictates behavior different from that based on a dispassionate weighing of political support. Ideology can prescribe actions that are economically unjustifiable and inefficient.

This chapter examines conservation ideology and how it affected Tucson policy. We conclude that the concept of conservation must include both development of and reduction in use of water resources, with the choice to increase supply or reduce demand being made on the basis of full consideration of economic, social, and environmental costs and benefits. This conception of conservation is a good deal more inclusive and balanced than previous, more conventional meanings that have either leaned heavily in the direction of growth and development or have opted for preservation and limits on use. As the Tucson case illustrates, the symbolic appeal of the idea of conservation has glossed over some very real differences in the concepts of values. This lack of coherence and consistency in meaning has resulted in actions taken in the name of conservation that may serve values very different from those held by other Tucson water users. Further, the narrow and partial perspectives about conservation adopted by participants in the Tucson case led to political difficulties for some of these individuals, and resulted in a more general failure to address the city's water problem effectively. Our intent as policy analysts is to draw on the lessons of the Tucson case that support a conception of conservation that can better guide municipal water policy.

Conserving What in Tucson?

Conservation is everywhere applauded, while there is nowhere a consensus on its meaning or on the conditions under which it should be practiced.[2] It is claimed that the Tucson experience "saved water," that is, reduced per capita use. Yet, because of the hydrologic cycle in which water is a continually renewing and reused resource, it is not always clear that saving water conserves anything.

It can be strongly argued that, under ordinary circumstances of surface-water supply, reduction of water use beyond normal cost-effectiveness makes no sense except in situations of drought.[3] As long as the available supply exceeds the quantity demanded, reduction in use by one user at one place or time simply shifts benefits to another user at another place or time, or, in the worst case, forgoes all benefits. Even in situations of drought, short-term conservation may introduce problems if it is used as a substitute for long-term capital investment. If total demand increases and there has been no expansion of the water system, when the next major drought occurs there may be no further slack in the system and no further savings can occur without undesirable economic repercussions.[4]

A different situation exists where the quantity of water demanded regularly exceeds the available supply, or where the long-run supply is finite, or where both conditions hold. Practicing "conservation" where it is relatively inexpensive to do so and transferring saved water to uses of higher value makes sense. Further, when nonrenewable supplies of groundwater are mined, conservation may be justified in the interest of future residents and the future economic base of the area. Tucson would seem to fit this situation. However, closer consideration of the motives of the major actors in Tucson water management since 1974 raises serious questions about the actual relevance of conservation to the events that occurred.

The most important rationale for price increases and public information programs in Tucson was financial conservation, not water conservation. According to the 1976 Carollo report, the city had fallen badly behind on capital improvements and needed to raise rates to pay for an expensive six-year capital-improvement program. Reducing peak use would help postpone the date when such improvements would become mandatory, while funds from higher rates accumulated. The capital conservation rationale was wholeheartedly endorsed by the staff of the Water Department as a solution to the department's financial difficulties. This same perspective was reinforced by consultants Black and Veatch in their 1977 study[5] and was endorsed by the new City Council. Financial management, rather than water conservation, has continued to motivate the city to the present time.

Further water development and use constitute the keystone of the Tucson "conservation" strategy. Development, however, will be staged to match the availability of funds, and water demand is managed so as not to outstrip currently available distribution systems. That conception of conservation, envisioning water development and use, is closely associated with one of the historic, if much criticized, strands of the conservation movement.

Shortcomings of Utilitarian Conservation

Gifford Pinchot, who credited himself with inventing the term *conservation*, founded and provided the intellectual underpinnings for the utilitarian strand of the conservation movement. Head of the U.S. Forest Service and one of Theodore Roosevelt's closest lieutenants, Pinchot believed that the first principle of conservation was development—the use of resources for the benefit of mankind.[6] The second principle of conservation was prevention of waste. Utilitarians believed in development with good housekeeping. Resources should not be squandered, or unnecessary refuse created. At the same time, failure to use resources, such as not harvesting mature trees or not harnessing rivers, was considered wasteful. The third principle of utilitarian conservation was management for the many rather than for the few. The formula for determining the common good was simple, if logically inconsistent: resources were to be developed for "the greatest good, for the greatest number, for the longest time."

As an intellectual force, the utilitarian or progressive strain of conservation had lost much of its strength by the mid-1950s. According to McConnell, the decline of the conservation movement was due primarily to the failure of its ideas.[7] Utilitarian conservation ideology did not provide an adequate guide for deciding difficult questions concerning natural resources. Some of these shortcomings are illustrated by the water conservation program as espoused by conservationists in Tucson.

Overemphasis on Development

Utilitarian conservation overemphasized development. While conservation was a gospel of efficiency, dictating that resource managers should maximize values relative to costs, the values of development were systematically overstated and the costs of development relatively ignored. Because utilitarians were development-oriented, the basic desirability of programs was rarely questioned. In fact, their analyses served to justify projects rather than to determine most efficient resource allocation from a general perspective. The negative environmental and social consequences of actions were seldom fully taken into account. The utilitarian philosophy as applied to water resources resulted in huge, expensive reclamation projects with severe environmental consequences. The aim of these projects was often to generate growth rather than to serve existing needs.

Sustained yield was a central tenet of Pinchot's forest management philosophy, but as the concept was implemented in U.S. Forest Service policy, it included social and commercial concerns that went well beyond ever-renewing forests. The Sustained Yield Act of 1944 was largely aimed at promoting the stability of forest industries and that of employment in communities dependent on the forest products. So construed, the concept of sustained yield as applied to the Tucson Basin means more than management to maintain the useful life of the aquifer by bringing recharge and withdrawals into balance. It also includes development of surface supplies to support the continuance of the building trades that have fed on population growth.[8]

Tucson's utilitarian conservation program overemphasizes growth and development. Low water prices were increased only in order to raise capital to develop. Water is priced at its historical average cost rather than at its marginal cost. As is discussed in the next chapter, average-cost pricing rather than marginal-cost pricing does not accurately signal the true economic or environmental cost of water-consuming growth. In fact, low water prices may actually encourage growth that generates pressures for further water development. In addition, Tucson's widely publicized success at water conservation may convey to investors and builders the idea that there will be no water problem in the future. Currently the Tucson Water Department is planning for a Tucson population of 1.8 million in 2035, up from only 0.5 million in 1983. The city water planners say this population will require reduction of per capita use to around 100 gallons a day, pumpage at extraordinary depth, and importation of very expensive water from the Central Arizona Project.[9] Yet the prevailing utilitarian conservation wisdom does not question the adequacy of water supply or suggest that finite supplies of water imply limiting population growth. A water policy less biased toward growth would at least consider the benefits and costs of nondevelopment and would price water at its full social cost.

Failure to Allocate Resources Efficiently

Utilitarian conservation failed to allocate resources efficiently. Gifford Pinchot's brand of conservation involved the scientific management of resources. Conservation was a gospel of efficiency dictating that resource managers should maximize the "good." Correct development decisions required the insights of technicians and experts using analytical tools.[10] Hence, utilitarian conservationists nominally believed in economic analysis and the careful weighing of benefits and costs. The federal agencies that have most strongly embodied the utilitarian ideals, the Forest Service and the Bureau of Reclamation, have long been committed to comprehensive planning and analysis and have required employees to prepare documents "justifying" proposed projects on economic grounds. Part of the analysis performed on projects, particularly those dealing with water resources where users were to pay back part of the cost of construction, involved findings of financial feasibility. The price set for the project goods and services reflected analysis of pay-back requirements and users' ability to pay.

Utilitarian conservationists were devoted to economic analysis, but their brand of analysis, as practiced, was incomplete and biased. Academic economists have argued that this kind of faulty economic analysis has led to inappropriate uses of price. Price is usually left out in assessments of water requirements, which are instead based on a vague accounting of water needs irrespective of what meeting those needs may cost. Prices are set to meet financial payback requirements and to attract political support. Instead, economists believe that price should reflect real social costs and lead to a more efficient allocation of the resource.[11] When the price of water reflects its true value—including social and environmental costs of development and the value forgone to future generations—the development of new supplies may not be justified because the market will operate to reallocate existing supplies to their highest-value uses.

Water allocation in Tucson through pricing policy presumably is the result of careful economic analysis. Yet, as is shown in the next chapter, there is flawed economic reasoning behind average-cost pricing and Tucson's application of "cost of service" concepts. Economic analysis of water pricing is performed as it is in Tucson because of the utilitarian commitment to serving water customers through further development of water resources. Pricing policies manage demand so that development can be paced in a financially convenient manner. Water rates are not set to maximize efficient allocations of resources.

In a larger context, water allocation in the entire Tucson Basin reflects the inefficiencies fostered by utilitarian conservation concepts. Irrigated agriculture is responsible for 54 percent of total consumptive use of water pumped from the aquifer, while municipal and industrial uses are responsible for only 30 percent. Tucson's proposed solution to overdrafting —that is, importing water through the Central Arizona Project—is typical of the historical utilitarian solution. In fairness it must be observed that the Groundwater Management Act passed by the state legislature in 1980 envisions reduction over time of agricultural pumping. Yet the inefficiencies in the allocation of water resources today are clear when Tucsonans are exhorted to save water despite the fact that most of the overdraft is created by agriculture.

Failure to Address Problems of Equity Successfully

The formula developed by Pinchot for determining how benefits from resource development are distributed was not operational. Herfindahl explains:

When this definition is read off rapidly—conservation is the use of natural resources for the greatest good of the greatest number for the longest time— the three superlatives have a delightful ring. If this is conservation, how could anyone be opposed to it? No one could be, of course, unless he stops to ask himself how three variables can be maximized at the same time. Imagine a father trying to distribute a bag of candy to his children so as to maximize the amount of candy received by each child who gets candy *and* the number of children receiving candy *and* the length of time the candy will be visible. [12]

Conceptual problems with utilitarian measures to obtain equity eventually led to challenges by groups disadvantaged by the agencies applying these measures. They argued convincingly that agencies were making value choices as to who should benefit without any guidance from society in general.

Tucson's conservation policy also involves equity decisions that clearly are vulnerable to criticism. Present and future water users in Tucson are asked to bear costs while housing developers and their newcomer customers gain benefits. Advertising campaigns ask water customers to use less water than they would otherwise buy so that new water development will not be required so soon and so that increases in

the water rates will not have to take place. Artifically low water demand and water rates encourage development that requires additional water use.

While criticism is not yet widespread, some residents (usually environmentalists, although academic economists tend to agree) do not believe Tucson's water conservation program is equitable. In an article in Tucson's *Arizona Daily Star* entitled "Beat the Peak Quackery," Charles Bowden argued:

What the Beat the Peak campaign will never do is address the basic problem: an expanding city on a shrinking water supply. Instead, Beat the Peak fosters growth by delaying costs that growth creates.

Imagine the aquifer as a bucket and each resident is busy slurping it up with a straw. Beat the Peak shares the straw with newcomers and never drinks between 4 and 8 or two days in a row. In this way, the need for new straws is postponed.

Of course this moment of hydrologic balance and economy will pass. As the city's population grows, the amount of water consumed will climb right back up where it was before Beat the Peak and high water rates came down the pike. The need for capital improvements will knock again at the door; the aquifer will continue its march beyond the human reach. This is the inevitable achievement of the Beat the Peak campaign: more people settled on less water. We will expand the problem and shrink the solution.

Beat the Peak is the latest effort of Tucson's boosters to avoid asking how large Tucson should become. Is this desert valley the right place to plant a huge human community? Without question it is the right place to temporarily peddle televisions, hustle automobiles, tout air conditioning and bid up real estate. But is it the right place for 1 million people for 50 years, 100 years, for good?

Let's really Beat the Peak. Instead of bragging about a reduced per capita consumption of water so that more people can be stuffed into this valley, let's turn on the faucets full blast.

Let's have that $45 million water [development] bill come due tomorrow so that folks settling here can have a taste of the future. Since there appears to be no appetite for limiting the size of the city or for bringing its water consumption in line with its water supply, let's get the water out of the ground and flowing north down the Santa Cruz before a megalopolis lands on our hands, courtesy of programs like Beat the Peak.

Finally, let's stop complaining about the groves south of the city because they use water. If they were gone, we'd have subdivisions guzzling down the aquifer instead of pecans.

So, come 4 o'clock in the afternoon, head out to the yard and get those hoses running. Let's have water roaring out of our yards and cascading over the curbs. Then crack a beer and turn on the television and wait for the moral strictures of Beat the Peak to be replaced by a new message: "Surf's Up."[13]

Long-run Problems Aggravated Rather Than Resolved

Utilitarian conservation in the long run aggravates rather than solves conservation problems. In the real world the goals of policies are seldom attained, and trying to solve problems creates new problems. Therefore, as Wildavsky argues, the real test of a policy is whether you like the new problem set better than the one you had previously—particularly, whether new problems are more tractable.[14] Critics of water reclamation policies, guided by utilitarian ideals, have argued that overbuilding our rivers like the Colorado has bequeathed enormous social, economic, and environmental problems. In his newspaper comment Bowden suggested that future problems in Tucson will be far less pleasant than today's water difficulties. With presently projected supplies, 1.8 million future Tucson residents will have to get along with 100 gallons per capita a day in the year 2035 if the aquifer is to be stabilized. Quality of life may well deteriorate. At some point population growth will have to stop, and it may be more difficult to construct a steady-state economy with 1.8 million than with 0.5 million residents. Should overdraft continue, eventually the aquifer will be completely depleted. Locating additional surface supplies or dismantling the city will pose no simple problem. It will be a curious turn of events if water conservation in Tucson leaves the future worse off, but that seems a clear possibility.

Shortcomings of Preservationist Conservation

While the prevailing conception of conservation in Tucson is utilitarian, some of the initiators of the Tucson water conservation movement had an alternative view. To the New Democrats on the City Council, the groundwater overdraft was just one indication that the City of Tucson was way out of balance with its desert environment. If Tucsonans were living in the proper harmony with nature, they would be able to pass along the city's aquifer resource to future generations unimpaired. For the New Democrats, conserving water was part of a larger strategy of controlling growth and maintaining environmental quality. The council majority believed that unplanned population growth was spoiling Tucson's living environment. New Democrats aimed to discourage urban sprawl and the infringement of housing projects into the foothills of the Santa Catalina and Tucson mountains. A planning commission responsible to the County Board of Supervisors, and particularly influenced by Ron Asta, set to work on a slope ordinance that would restrict building on hillsides. The lift-charge system of water rates, requiring foothill residents to pay more, nicely complemented the intent of the slope ordinance. The New Democrats were expressing standard environmentalist ideology that the costs of environmental damage should be internalized—that is, those responsible ought to pay the price. Directly experiencing costs forces people into greater compatibility with nature. System development charges to newcomers who settle at the edge of the sprawling city and require new water hookups could be justified under this rubric. So, too, increasing block rates could be supported, because these rates charged more to users who were perceived as taking more than their fair share of a scarce resource.

The roots of the unity-with-nature philosophy can be traced to the preservationists. This strain of conservation, like the utilitarian strain dates back to the Theodore Roosevelt era, but advocates the preservation rather than development of resources. Expressed through the writings of men like George Perkins Marsh (his *Man and Nature* was published in 1864) and John Muir (*Our National Parks* appeared in 1901), preservationists believed that leaving resources in their natural state was often the highest purpose. Believing that all the wild things of creation had an importance beyond their usefulness to man, preservationists fought to preserve and protect their integrity. [15] In a classic preservationist battle, John Muir and his followers opposed the City of San Francisco and the utilitarians over the development of the Hetch Hetchy Valley. For the preservationists, the highest use of a valley just as beautiful as Yosemite was not— as long as other alternatives for urban water supply existed—to become submerged beneath a reservoir. While not opposed to the development of places of lesser scenic quality, the preservationists were skeptical of the benefits of development in general. Given a choice, they preferred to ration present use of resources to ensure that future generations would have at their command stock that was not substantially depleted.

There is not a perfect relationship between preservationist ideology and modern-day environmentalists, including the New Democrats of environmentalist persuasion on the Tucson City Council. Even the Sierra Club, which traces its roots to John Muir himself, has adopted other concerns, including pollution and the urban environment. The New Democrats differed one from another in terms of environmentalist credentials, and as politicians and public officials they were motivated by a good deal more than ideology. At the same time, when it comes to matters of conservation, environmentalists' heritage biases them against development and toward restricting use. Such a reaction, driven by preservationist ideology, may lead to serious miscalculations and mistakes.

Saving of Resources Even When Benefits Are Doubtful

As a general position, preservationists prefer not using resources and leaving them in their natural state. Human infringement on nature tends to be seen as an intrusion. Because of the advance of technology and rapid population growth, preservationists fear that man will destroy the habitat upon which all living things are dependent. The adverse consequences of changes may be irreversible and become recognized too late to change actions.

Preservationists tend to take a negative view of new resource development and see using more as undesirable behavior. Nature in its natural state is best, they think; the desert should look like desert, and persons who create a "jungle" in a desert should be made to sacrifice monetarily. It is also unnatural to build houses high in the foothills, especially when space is available on flat lands. Those who choose the mountain vista and cause more rapid erosion should be penalized.

As a strategy to make the public sensitive to its natural surroundings, and as a means to inculcate the value of frugality, the preachment of restricting use may have an important function. However, in many situations, following its strictures delivers few real benefits.

Water often is a renewable resource with flow characteristics. Runoff from precipitation flows into river systems and is diverted and used nonconsumptively or only partially consumptively, and wastewater is returned to the stream. In its course to the ocean, water may be reused numbers of times, providing a wide variety of benefits. Deciding not to divert water and forgoing particular benefits at any point along the stream can be a sacrifice that makes no one better off, unless there is some sort of shortage downstream or instream. To be worthwhile even in this case, the downstream use would have to be of higher value. In the case of groundwater, recharge occurs, and nonuse per se is hard to justify. Even the mining of groundwater beyond what is naturally replenished may not necessarily be bad. It can be argued that leaving the same amount of water in the aquifer forever is like leaving money in a bank forever without collecting interest. Neither action produces any benefit.

In circumstances where restricting the use of water operates to facilitate long-run growth and development, saving water may in fact be contrary to preservationist values to protect the natural environment. In Tucson, beating the peak contributes to orderly water development and to the growth of the city. It is not certain whether the New Democrats were aware that residential water conservation served the interests of developers and real estate interests. They may have thought that as long as savings were occurring, the rationale did not matter. The motives of the utilitarian conservationists, to postpone capital investment, may have been seen as additive or reinforcing.

Further, the New Democrats' preservationist focus on municipal water savings did not address the general issue of overall water management in the Tucson Basin. Had long-term preservation of the aquifer been the major concern, the council majority also would have had to consider the larger withdrawals by agriculture. Perhaps the New Democrats did not think that they had to concern themselves with water use beyond the city limits so long as their policies were promoting a morally correct conserving behavior among residents.

Bias Against Economic Analysis

Preservationist conservation is biased against economic analysis as a basis of decision. Preservationists have thought that the worth of resources in their natural state is better expressed through pictures and literature than by direct analysis. Many preservationists have doubted economists' ability ever to evaluate aesthetics and natural habitats. For instance, Aldo Leopold, in *A Sand County Almanac*, pokes mild fun at economists who use statistics to characterize his Wisconsin farm as poor, yet lack any understanding of its value for such unique and irreplaceable purposes as the stage for the male woodcock to perform his mating dance.[16] Preservationists are also critical of economists' time horizons, which in their view are entirely too limited. Rather than basing analysis on discount rates that systematically discount the present value of a future stream of benefits or costs, preservationists believe that correct analytics should take into account the ultimate exhaustibility of the earth's resources. Ways should be found to ration the use by present generations to ensure that future generations will have at their command a stock that is not substantially depleted. Thus, preservationists tend to be favorable to such concepts as "safe yield" of groundwater, which balances withdrawals and recharge, even though this concept may not make economic sense to present users.

While preservationists have not generally espoused economic analysis as an appropriate method of evaluating their own preferences for restricted use, they have become skillful in using economics to discredit projects to which they object. Preservationists are also interested in pricing strategies that will discourage or stop the use of natural resources. For example, environmental groups have applauded higher user charges for commercial barges using the nation's system of locks, canals, and dredged channels.[17] They have also supported the proposition that states and localities pay larger front-end costs of federal water development projects.[18] The issue here goes beyond the economists' notion that users should pay. Instead, prices are viewed by some environmentalists in terms separate from their notions of worth. As Wildavsky remarked about this phenomenon:

> Some things, after all, must be sacred. Environmentalists are trying to move the boundaries by which men distinguish between the profane—money, the economic calculus—and the sacred—man's relation to nature. God is not dead: only immanent in nature . . . confusion enters because the transactions occur between two worlds, so that homage must be paid to the old economic costs and benefits whereas choices are predicated on quite different environmental values.[19]

Some of the rhetoric of the council majority suggests that they did believe that using lots of water in Tucson was profane. They voted to ban all outdoor use on Wednesdays and prescribed alternate-day watering. Had the New Democrats been more sensitive to economic analysis, they would not have so readily supported increasing block rates. As we will explain, increasing block rates cannot be justified by economic reasoning. Rather, the underlying justification is preservation morality—large users should pay more for each increment of water than should smaller users. Without an economic rationale, New Democrats were vulnerable to the charge of having been discriminatory, in the end a politically costly mistake. Further, economic analysis would have suggested marginal-cost pricing through which residents can get clear signals of the costs of water use that requires additional water development.

Ideology Often Set Above Political Feasibility

Preservationists note that humans are only one species among the earth's living creatures. Ecological balance is more important than human preferences. Therefore, the logic goes, preservationists should do what is right even if it is unpopular. There is a self-righteousness about environmentalists that is characteristic of true believers of all kinds.

The case study reported in chapter 2 illustrates a surprising disregard by some elected officials for political feasibility. The discussion of politics in chapter 6 seeks an explanation for the council majority's failure to read political signals. Briefly stated though, some elected officials gave higher priority to serving their moral values than to the job of mirroring public attitudes. The council majority was not successful in communicating to voters its belief in the higher morality. The council's actions were perceived as high-handed rather than as courageous. At least part of the New Democrats' leadership problem related to the inconsistencies in the perceived meaning of conservation.

Learning From Conservation Preachments

Giandomenico Majone has likened policy analysis to generalized jurisprudence. The analyst's task is to examine the clarity, consistency, and internal logic of the concepts that undergird public action.[20] In this chapter we have subjected the concept of water conservation to such an examination.

Conservation of natural resources has long been a cause with great symbolic appeal, yet historical research of the use of the term reveals little consensus as to its meaning.[21] Differences began at the birth of the conservation movement and have continued to the present. The two main branches of conservationists—preservationists and utilitarians—have differed about the implications of conservation for growth and devel-opment. Further, various conceptions of conservation relate differently to economic analysis and to whether and how pricing should be used as an instrument to achieve conservation. The difficulties that have arisen during various attempts to implement conservation as a policy have occurred most often because the brand of conservation being applied was partial and limited and therefore devisive and ineffective.

A number of actions were taken in Tucson in the name of conservation, yet participants supporting reductions in water use had very different values about the desirability of growth and development versus preservation of the desert environment. The focus of concern was upon immediately restricting use, disregarding the long-term implications of cutbacks and the question of which interests might ultimately benefit or lose. For the utilitarians, conservation meant relief from the Water Department's financial crisis, and security that the city could continue to grow without the specter of water shortages discouraging developers. For those of a more preservationist persuasion, reduced water use was a step in the direction of bringing the city into harmony with its desert environment. The future of Tucson now seems to be one of very rapid growth, with water conservation to facilitate that development. Tucson appears to have chosen the flaws of utilitarians over those of preservationists. Yet, the growth issue has never been joined with the water issues because the symbol of conservation smooths over differences.

The preachment of conservation better serves water policy when it dictates open and rational choice about growth and development on the basis of full information about costs and benefits and who bears them. The concept of conservation needs to integrate the diverse meanings of the past and to incorporate a consistent pricing strategy effectively. The Tucson case suggests that conservation can be served by both development and reduction in use. A reasonable conservation strategy would be one in which water demand and supply management are integrated so that economic and environmental benefits of water policy exceed economic and environmental costs.[22] New resources development is not justifiable just because there is demand and funding for projects. Rather, benefits must exceed costs in a full accounting including social and environmental factors. The full price of new development must be borne by beneficiaries in an equitable manner. By the same reasoning, nonuse cannot be justified as conservation. Self-sacrifice as an end in itself is not logical or politically meaningful. Environmentalists no longer need to rely on symbolic actions aimed at gaining public attention. After more than a decade on the public agenda, the skepticism about both development and preservationist values are widely shared.[23] Ways are needed to implement policies reflecting this concern. This situation has led many contemporary environmentalists to become increasingly more amenable to quantification and compromise. In the words of Wildavsky: "Environmentalism now uses technology. An instrumental device, cost-benefit analysis appears in the service of an expressive ideology. The pure symbol will not go down, and so it is overlaid with ameliorative measures, stripping it of its severity, made a matter of more or less rather than all or nothing."[24]

The Tucson case illustrates the willingness of conservationists of both strains to consider economic concepts such as pricing on the basis of cost of service. The reasoning behind the use of these concepts was flawed. The difficulty, however, does not lie entirely with the conservationists. Noted economist Steve Hanke admits:

All of the blame for this new set of events [in which inappropriate new pricing strategies have been adopted] . . . cannot be placed on the shoulders of the environmentalists who pursue the goal of conservation. Moreover, it cannot be placed on the shoulders of the industry, which accepts the environmentalists' faulty arguments. Economists are largely to blame. If economists want to stop experiencing nightmares, they must begin to apply their concepts and they must communicate with those persons responsible for implementing water-rate policy.[25]

The next chapter takes up the task of an applied analysis of municipal water pricing.

Notes

1. Alfred L. Kroeber, *Cultural and Natural Areas of Native North America* (Berkeley, University of California Press, 1939).

2. Dean Mann, "Institutional Framework for Agricultural Water Conservation and Reallocation in the West: A Policy Analysis," in Gary D. Weatherford, ed., *Water and Agriculture in the Western U.S.: Conservation, Reallocation, and Markets* (Boulder, Colo., Westview Press, 1982) p. 11; and Orris C. Herfindahl and Allen V. Kneese, *Quality of the Environment: An Economic Approach to Some Problems in Using Land, Water and Air* (Baltimore, Md., Johns Hopkins University Press for Resources for the Future, 1965) pp. 229–236.

3. William Whipple, Jr., "An Economic Analysis of Water Conservation Policy," *Water Resources Bulletin* October (1981) p. 818.

4. Ibid.

5. Black and Veach, Consulting Engineers, *Report on Capital Improvement Program, Revenue Requirements, Water Rates*. Submitted to City of Tucson Water Utility (Kansas City, Mo., June 28, 1977).

6. Grant McConnell, "The Conservation Movement—Past and Present," *Western Political Quarterly* September (1954) pp. 463–478.

7. Ibid., p. 470.

8. Paul W. Bedard and Paul N. Ylvisaker, *The Flagstaff Federal Sustained Yield Unit* (University, Ala., University of Alabama Press, 1957).

9. Interview with Stephen Davis, Tucson Water Department, Tucson, Ariz., April 17, 1982.

10. Samuel P. Hays, *Conservation and the Gospel of Efficiency* (Cambridge, Mass., Harvard University Press, 1959).

11. Maurice M. Kelso, William E. Martin, and Lawrence E. Mack, *Water Supplies and Economic Growth in Arid Environment: An Arizona Case Study* (Tucson, University of Arizona Press, 1973) pp. 224–225.

12. Orris C. Herfindahl, "What is Conservation?" in D. L. Thompson, ed., *Politics, Policy and Natural Resources* (New York, Free Press, 1972) p. 172.

13. Charles Bowden, "Beat the Peak Quackery," *Arizona Daily Star*, August 2, 1981, p. C-3.

14. Aaron Wildavsky, *Speaking Truth to Power* (Boston, Little, Brown, 1979).

15. McConnell, "The Conservation Movement"; George Perkins Marsh, *Man and Nature* (New York, Charles Scribner, 1864); and John Muir, *Our National Parks* (Boston and New York, Houghton Mifflin, 1901).

16. Aldo Leopold, *A Sand County Almanac* (New York, Oxford University Press, 1962).

17. See particularly the *Coalition for Water Project Review Newsletter* of April 30, June 30, and July 15, 1981 (Washington, D.C., Coalition for Water Project Review).

18. James Q. Wilson, *Political Organization* (New York, Basic Books, 1973).

19. Wildavsky, *Speaking Truth*, p. 191.

20. Giandomenico Majone, "Technical Assessment and Policy Analysis," *Policy Sciences* vol. 9 (1977) pp. 173–175.

21. For example, see Herfindahl, "What is Conservation?" p. 172; See also Leonard Shabman, "Water Conservation as a Basis for Reforming Water Resource Programs." Paper presented at the American Water Resources Association Meetings, Las Vegas, Nev., September 24–27, 1979; and Mann, "Institutional Framework."

22. Shabman, "Water Conservation," p. 10.

23. Council on Environmental Quality, Environmental Protection Agency, U.S. Department of Energy, U.S. Department of Agriculture, *Public Opinion on Environmental Issues* (Washington, D.C., Government Printing Office, 1980).

24. Wildavsky, *Speaking Truth*, p. 200.

25. Steve H. Hanke, "Pricing as a Conservation Tool: An Economist's Dream Come True?" in David Holtz and Scott Sebastian, eds., *Municipal Water Systems* (Bloomington, Indiana University Press, 1978) p. 238.

4

Pricing Municipal Water

We have argued that the preachments for water conservation must be integrated into a logical, consistent system if they are to be effective in balancing a long-run water supply with long-run water demand. Traditionally, municipal water utilities have focused only on providing an "adequate" supply at the lowest possible cost. Individuals' demands for water have been assumed given, aggregate demand for water has simply been a function of population, and the quantity of water demanded has been assumed to be almost unrelated to the water's price.

Economic theory provides a rational, integrated system for examining questions of both supply and demand. Economics prescribes that a rational goal is to develop water supplies only if the benefits to consumers are equal to or greater than the cost of the development. Further, in terms of equity, and to assure that the decision to develop is based on serious consideration, it should be the beneficiaries themselves who pay the costs. To satisfy the conservation ideologies, cost must be defined to include not only direct development costs, but also indirect costs such as those of environmental degradation. Benefits are defined as the maximum amount the consumers would pay to obtain both the newly de-

veloped water and any indirect benefits the project might generate.

Thus, to ensure that water is developed only if consumers really desire the additional water *as expressed by their willingness to pay for it*, all water should be priced at the cost of developing an additional unit of supply. This system, termed *marginal-cost pricing*, is the economists' view of rational conservation behavior. It is related to utilitarian conservation in that it is not antidevelopment. However, the system focuses rigorously on the costs as well as the benefits of development.

The first six sections of this chapter describe the economics of water utility rate making, evaluate the traditional procedures, and suggest a new procedure that could be used if a utility accepted marginal-cost pricing. This analysis is relatively technical. Many readers may wish to skip the technical discussion and take up the chapter at the section entitled "Tucson Utility Rate Schedules." This last part of the chapter describes how water has been priced in Tucson, and compares the results to marginal-cost pricing in light of its effect on conservation behavior. This analysis is of the supply side of the water management issue. Chapter 5 then examines Tucson consumers' willingness to pay for water—the demand side of the issue. Our goal is to examine the importance of water price for consumer behavior, relative to preachments, politics, and practices.

Economics of Utility Rate Making

In recent years interest in examining the pricing policies of public water utilities has increased nationwide. Continuing inflation, combined with a gradual increase in the real cost of supplying piped water, has caused water utilities to raise their rates with increasing frequency in an attempt to cover their costs. The need to revise water rates regularly has led the water supply industry to take more interest in rate-making practice. There is a growing feeling in the industry that water rates should be designed to recover the costs of providing water by charging customers in accordance with how they contribute to the costs.

Frequent increases in water rates also have led the public to take more interest in water utilities' pricing policies. Public interest groups, especially those of preservationist bent, have asserted that the traditional form of water rate subsidizes customers using large amounts of water at the expense of more frugal customers and encourages excessive consumption of water.

This new interest in pricing policies has led a number of water utilities across the United States to introduce innovative rate schedules. The City of Tucson Water Department is among a small number of departments that have introduced increasing multiple-part tariffs in which the price of a unit of water increases as a customer's consumption exceeds various levels of use within each billing period. As another example, the Fairfax County Water Authority of Virginia has introduced a rate schedule in which customers are charged extra for summertime use that exceeds 130 percent of their average winter use. Generally speaking, these innovative rate schedules have been designed with the intention of charging customers for water in a way that closely reflects the cost of serving these customers, although to some extent their purpose has been punitive, as was discussed in chapter 3. This approach to rate making is generally

believed to price water in a way that both allocates costs fairly among customers and encourages water conservation.

The growing feeling in the water supply industry that costs should be closely reflected by rates certainly is welcome. However, the procedures that currently are used in the industry to determine water rates are not based upon sound economic analysis. The deficiencies in these procedures seriously limit the extent to which water rates can achieve their objective of charging customers in accordance with the costs of providing service. This section discusses the problems of the traditional approach to rate making and describes how water rates can be designed using a logically consistent economic approach.

The water supply industry, like most public utilities, is characterized by high fixed costs and has generally had large economies of scale. Typically, at least half of the cost of supplying piped water results from the expense of financing the construction of the works needed to procure, transmit, treat, and distribute water. In most cases, the costs of the works used to procure water constitute a fairly small part of a water utility's total investment. Howe and Linaweaver suggest that, in the eastern United States, about 30 percent of a typical water utility's investment is in works for the procurement of water, a further 20 percent of the investment is in transmission and treatment plant, and the remaining 50 percent of the investment is in distribution plant.[1] The findings of a survey of water utilities reported by Hanke indicate the typical investment in procurement works to be less than 10 percent of a water utility's total investment, the investment in transmission and treatment plant about 45 percent of total investment, and the investment in distribution plant about 35 percent of total investment.[2]

The division of a water utility's investment into works for the procurement of water, transmission and treatment plant, and distribution plant is of particular interest because these three groups of plant are designed to meet three different kinds of demand. The works for the procurement of water are designed to meet the year-round average demand. Transmission pipelines, pumping stations, and treatment plants are usually designed to meet demand over the peak day of the year. Distribution mains, pressure tanks, and service reservoirs are designed to meet demand during the peak hour. Since the distribution plant typically accounts for a large proportion of a utility's investment, the amount of water demanded during the peak hour has a strong influence on the cost of constructing and operating a water utility.

The technology of the water supply industry is such that there has been a strong tendency for there to be net economies of scale. For example, the cost of each unit of capacity of a main or pipeline decreases as the capacity of the pipe increases. The costs of a pipe in terms of its capacity can be approximated by a function of the form

$$C = kQ^a$$

where C is the cost of the pipeline, k is a constant, and Q is the capacity of the pipeline; the exponent a is a dimensionless number, less than 1, known as the economy-of-scale index. If the exponent a is less than 1, costs increase at a decreasing rate as capacity increases. Scareto[3] reports studies indicating that the economy-of-scale index for pipelines can be as low as 0.58. Since the major cost of a distribution system is the expense of laying the extensive network of mains that convey water from the transmission pipeline and service reservoirs to individual customers' service connections, the cost of each unit of delivery capacity of a distribution system will fall as capacity increases. Similarly, transmission pipelines show strong economies of scale with respect to their delivery capacity.

The works for the procurement of water usually show economies of scale until the physical limits of the source of water are approached. Since a utility normally will develop the least expensive source of water first, there will be economies of scale until water consumption exceeds the amount of water that can be taken from the utility's least expensive source. As consumption exceeds the total quantity of the first source of water, costs will rise sharply, and then gradually decline again as economies of scale in the second source of water are realized. In this situation, where there are economies of scale in some respects but diseconomies of scale in other respects, it is particularly important to distinguish between the utility's average and marginal costs.

Consider the case of a water utility that has three sources of water: the first source, which is the cheapest, can supply up to 30,000 acre-feet annually; the other two sources, which are somewhat more expensive, can each supply up to 10,000 acre-feet annually. The total costs of developing these three sources of water are as follows:

$$
\begin{align}
C_1 &= 30q_1 - \tfrac{1}{3}q_1^2 \quad \text{for } 0 < q_1 < 30 \quad (1)\\
C_2 &= 35q_2 - \tfrac{1}{2}q_2^2 \quad \text{for } 0 < q_2 < 10 \quad (2)\\
C_3 &= 40q_3 - \tfrac{1}{2}q_3^2 \quad \text{for } 0 < q_3 < 10 \quad (3)
\end{align}
$$

where q_1, q_2, and q_3 are the annual quantities of each of the three sources in thousands of acre-feet, and C_1, C_2, and C_3 are the total annual costs in thousands of dollars of developing these three sources of water. Within each source, total costs increase at a decreasing rate, indicating economies of scale.

The marginal cost of taking water from each of the three sources—that is, the cost of developing each source to yield an additional unit of water—is found by differentiating each cost function with respect to quantity. The marginal costs are:

$$
\begin{align}
MC_1 &= 30 - \tfrac{2}{3}q_1 \quad \text{for } 0 < q_1 < 30 \quad (1')\\
MC_2 &= 35 - q_2 \quad \text{for } 0 < q_2 < 10 \quad (2')\\
MC_3 &= 40 - q_3 \quad \text{for } 0 < q_3 < 10 \quad (3')
\end{align}
$$

where MC_1, MC_2, and MC_3 are the costs of each additional acre-foot of water from each of the three sources.

The water utility presumably will wish to obtain its water as cheaply as possible, and so first will develop source 1, turning to source 2 when consumption exceeds 30,000 acre-feet per year, and then to source 3 should consumption exceed 40,000 acre-feet per year. The average cost of an acre-foot of water is the total cost divided by the quantity of water delivered. Thus, for:

up to 30,000 acre-feet

$$AC_1 = (30q_1 - \tfrac{1}{3}q_1^2) / q_1 \quad (1'')$$

between 30,000 and 40,000 acre-feet

$$AC_2 = (600 + 35q_2 - \tfrac{1}{2}q_2^2)/(30 + q_2) \quad (2'')$$

between 40,000 and 50,000 acre-feet

$$AC_3 = (900 + 40q_3$$
$$- \tfrac{1}{2}q_3^2)/(40 + q_3) \quad (3'')$$

The marginal and average costs of providing water service will vary with consumption, as shown in figure 4-1. As consumption increases from 0 to 30,000 acre-feet per year, the average cost of water, found by dividing the cost of source 1 by the total quantity delivered, falls from $30 per acre-foot to $20 per acre-foot. The cost of each additional unit of capacity also falls as consumption increases, reaching $10 per acre-foot when source 1 is fully developed. Because there is an overall economy of scale in this range of consumption, marginal costs are less than average costs. When consumption exceeds 30,000 acre-feet per year, the costs of each additional unit of water jumps to $35 per acre-foot as the more expensive water from source 2 is used. Marginal costs then fall as the economies of scale of source 2 are realized, but marginal costs are above average costs because net diseconomies of scale have occurred. Marginal costs again rise sharply when consumption reaches 40,000 acre-feet per year and water from source 3 has to be used.

Average costs begin to increase as water from source 2 is used, but because the average cost reflects the cost of all water used, including the cheaper water from source 1, average costs remain lower than marginal costs. Despite the economies of scale in the individual sources of water, an overall diseconomy of scale exists when total consumption exceeds 30,000 acre-feet per year. At these levels of consumption, marginal costs exceed average costs, and so the cost of obtaining additional water exceeds the average cost of the utility's supplies.

Many water utilities have reached the limit of their cheaper sources of water and are now encountering net diseconomies of scale. In the past, the cost of supplying water nearly always fell as consumption grew. But as the cheaper sources of water have become fully developed, many water utilities have had to turn to more distant, more costly sources of water. The cost of obtaining and delivering additional water now exceeds the average cost of the quantity of water currently being delivered. Such is the case in Tucson. Well fields were originally developed in areas close to the city with shallow pumping depths. Now well fields are outside the city, across the Tucson Mountains in the Avra Valley, where depth to water is greater. Future surface-water supplies from the Central Arizona Project will be much more expensive than currently developed groundwater supplies. The average cost of water is rising, but is much lower than the marginal cost.

The Demand for Water

In many cities water consumption increases markedly during the summer as large amounts of water are used for watering lawns and for feeding evaporative coolers. Since the indoor use of water for general sanitary purposes does not vary over the course of the year, this seasonal outdoor use largely determines the extent of demand over the peak days of the year.[4] In Tucson, Arizona, the peak-day demand on the city's water system is about twice the year-round average daily demand; in Denver, Colorado, the peak-day demand is 1.7 times the average daily demand; in Fairfax County, Virginia, the peak-day demand is 1.6 times the average daily demand.[5] Generally speaking, the ratio of peak-day damand to average-day demand for American water utilities ranges

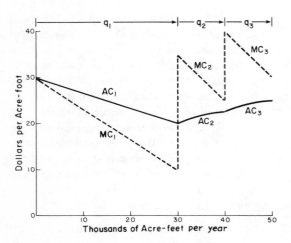

Figure 4-1. Economies and diseconomies of scale for a water utility (hypothetical).

from 1:1 up to 4:1, depending on the local climatic conditions, the size of the particular system, and the diversity of incomes and preferences of customers.

The demand for water also shows a marked cycle during each day. During the night, consumption is very low. It increases sharply in the morning to a morning peak and then declines somewhat until late afternoon. Consumption rises to an evening peak and then gradually declines to its low nighttime level. During the summer the use of water for sprinkling and coolers is highest in the late afternoon and early evening, resulting in very high peak demands on summer afternoons. Again, the ratio of the peak-hour demand to the annual average demand depends upon local climatic conditions and upon the diversity of the water system. In both Tucson and Denver, the peak-hour demand is 3.5 times the average daily demand.[6] Hanke suggests that, typically, peak-hour demand is 2.5 times the average daily demand on a water system.[7]

Generally speaking, the demand for water by a utility's individual customers shows much more variation than does the total demand on the system. Hanke reports that the peak-hour demand by individual residential customers is typically five times their average daily demand.[8] It is conventional wisdom in the water supply industry that consumption by commercial and industrial users peaks less strongly than does consumption by residential customers. However, the relationship of a commercial customer's peak demand to that customer's average demand will depend on the customer's particular kind of business. A residential customer's peak demand for water is largely determined by the extent of outdoor water use for sprinkling and cooling, suggesting that the peak-day-to-average-day ratio for an individual residential customer would increase as that customer's average consumption increases. Similarly, since a customer's peak-hour demand results from the superposition of early evening indoor demand with peak use for sprinkling and cooling, it also seems reasonable to assume that the ratio of peak-hour demand to average-day-demand increases as average consumption increases. Water rates recommended to the City of Tucson Water

Department by John Carollo Engineers in 1976 were calculated by assuming that customers' peak demands were related to their average demands according to the scheme shown in table 4-1.

The American Water Works Association's Rate-making Procedures

The water supply industry's traditional rate-making procedures are described in the American Water Works Association's (AWWA's) manual, *Water Rates*.[9] These procedures are intended to give schedules of water rates that charge customers for water in a manner reflecting the cost of serving these customers. Traditionally the AWWA's procedures have been applied to give rate schedules consisting of decreasing multiple-part tariffs, where the price of a unit of water falls as a customer's consumption exceeds certain levels each billing period. However, in recent years these procedures have been applied to give novel forms of rate schedules, such as increasing multiple-part tariffs, or schedules in which customers are charged extra for summertime consumption exceeding their average winter consumption.

The objectives of the AWWA's rate-making procedures cannot be faulted. Schedules of water rates that charge customers in accordance with the cost of service would be efficient from the economic point of view in that the price of a unit of water would be equal to the cost of the resources used to obtain and deliver that water, and they would be equitable in that no customer would be required to subsidize any other customer. However, the AWWA's rate-making procedures would achieve their intended objective only by accident, since their conceptual foundations are unsound. They are not based on marginal-cost theory. They cannot be, since costs are improperly defined and arbitrarily allocated. The existence of economies of scale has been a historical reason for avoiding marginal-cost pricing in the past, but it is unlikely to be a good reason in the future.

The definitional problem is that the ''costs'' used in the procedures are not costs in the economic sense, but are the water utility's revenue

Table 4-1. Peaking Factors for Tucson, 1976

| Class of customer | Daily and hourly peak demand as percentage of average daily demand Monthly consumption | | | | | | | |
| | Up to 600 ft^3 | | 600 to 1,001 ft^3 | | 1,100 to 2,000 ft^3 | | Over 2,000 ft^3 | |
	Peak day	Peak hour	Peak day	Peak hour	Peak day	Peak hour	Peak day	Peak hour
Single-family residences	200	280	250	450	300	590	340	740
Apartments and townhouses	200	280	200	300	250	400	300	590
Commercial	200	280	200	280	200	280	200	280

Source: Adapted from John Carollo Engineers, "Report Submitted to the Metropolitan Utilities Management Agency," final report (Tucson, Ariz., March 11, 1976).

requirements. Although a utility's revenue requirements are costs in the sense that they represent funds that must be paid to bondholders, wages that must be paid to employees, and so on, these revenue requirements do not directly represent the cost of providing service. The amount of revenue that a utility needs to raise each year depends not only upon the utility's current operating, maintenance, and development costs, but also upon financial obligations resulting from past capital expenditures. Revenue requirements represent a utility's historical costs. They clearly are not the marginal costs of providing a customer with an additional unit of water.

Consider the situation represented by the cost curves shown in figure 4-1. Suppose that the utility's three sources of water are alternative well fields and that the cost of developing these well fields consists solely of the cost of installing the wells and pumps and the cost of the pipelines needed to deliver the water to the utility's distribution system. Assume, for simplicity, that the utility finances its capital expenses by bonds issued in perpetuity at an interest rate of 10 percent. Thus, the annual costs of developing the three sources of water as given in equations (1), (2), and (3) are the annual interest payments resulting from the capital expenditure needed to develop each source to a particular annual yield. For example, the capital cost of developing source 1 to yield 20,000 acre-feet annually is such that the annual interest payments are $467,000 [$30(20) − \frac{1}{3}(20)^2 = 467$]. Similarly, the capital cost of developing source 2 to yield 10,000 acre-feet per year is such that the annual interest payments are $300,000.

Suppose now that water consumption in the city served by the utility is 30,000 acre-feet per

year, and that the utility has developed source 1 to its full capacity of 30,000 acre-feet per year. The annual interest payments resulting from the cost of developing source 1 to this capacity are $600,000, so that the average cost of water obtained from source 1 is $20 per acre-foot. The managers of the utility expect consumption to increase to 40,000 acre-feet per year, and they propose to meet this demand by developing source 2 to yield 10,000 acre-feet per year. The annual interest payments resulting from this development will be $300,000, making the utility's total interest payments $900,000 per year. Thus, the average cost of the 40,000 acre-feet to be obtained annually from sources 1 and 2 would be $22.50 per acre-foot. However, this cost of $22.50 is the utility's historical average cost and does not represent the real cost (the marginal cost) of developing the utility's sources of water to meet an annual demand of 40,000 acre-feet per year. The cost imposed upon the utility as consumption increases from 30,000 acre-feet per year to 40,000 acre-feet per year is $300,000 per year, which means that the average cost of the last 10,000 acre-feet developed is $30 per acre-foot. This cost—the cost of the water from the marginal source—is the cost incurred as consumption increases to 40,000 acre-feet per year; it would be avoided if consumption were to remain at 30,000 acre-feet per year. In this sense, the marginal cost of water represents the real cost of meeting a particular level of consumption.

Water rates based on a utility's historical average costs will raise sufficient revenue to meet the utility's annual interest payments, but will not represent the real cost of developing the water. If the utility developed source 2 and then computed its rates to reflect its new revenue requirements, the result would be that water costing up to $30 an acre-foot to obtain would be sold at a price of only $22.50 per acre-foot. It could be that this water is not worth $30 per acre-foot to the utility's customers. It is possible that these customers might choose to curtail their consumption, obviating the need for the utility to develop source 2, rather than to buy water at the real marginal cost of $30 per acre-foot. In such cases, the use of rate schedules based on average costs rather than on marginal costs will result in overbuilding and in the development of water that costs more to supply than it is worth to consumers.

As discussed in the preceding section, marginal costs are less than average costs when economies of scale exist, but become greater than average costs if diseconomies of scale are encountered. In the past the water supply industry has generally been a decreasing-cost industry with marginal costs being less than average costs. In that situation, rate schedules based on marginal costs will not raise sufficient revenue to meet a utility's financial obligations, and the utility cannot be self-supporting. Thus, it is not surprising that utility managers have seen average cost per unit of water delivered as being representative of the real cost of providing service and have used these historical costs as the basis of their rate schedules. However, when net diseconomies of scale are encountered, the marginal cost per unit of water delivered will exceed the average cost, enabling a utility using marginal-cost pricing to be self-supporting. In these circumstances there is little justification for continuing the practice of basing water rates on the utility's historical average cost per unit of water.

Another shortcoming of the AWWA's procedures is in the way total costs are allocated. The total costs of financing and operating the water system are allocated among the various functions, and then these allocated costs are used to set prices in the various use blocks of a multiple-part tariff. Two different ways of allocating costs between functions are suggested: the Commodity/Demand method and the Base/Extra Capacity method. In the Commodity/Demand method, the utility's costs are divided into Commodity Costs, Peak-day Demand Costs,

Peak-hour Demand Costs, and Customer Costs. Commodity Costs are the costs of the actual delivery of water; Peak-day and Peak-hour Demand Costs are the costs of providing the extra capacity needed to meet the peak demands; and Customer Costs are the costs of hookup, meter reading, and administration. In the Base/Extra Capacity method, costs are divided into Base Capacity Costs, Peak-day Extra Capacity Costs, Peak-hour Extra Capacity Costs, and Customer Costs. In this method, Base Capacity Costs are the costs of the actual delivery of water plus the cost of providing the capacity needed to meet the year-round average demand; Peak-hour and Peak-day Extra Capacity Costs are the costs of providing the extra capacity needed to meet the peak-hour and peak-day demands; and Customer Costs are, again, the costs of providing customers with service. Peaking costs are important in both methods. Thus, it is obvious that Tucson's Beat the Peak campaign is a method of financial conservation.

Neither of these two methods of allocating costs develops the real costs of providing for the year-round average demand, the demand over the peak day, and so on. Both methods take the utility's total costs (or more precisely, the utility's revenue requirements) and allocate these costs to one or another of the specified functions. But not all of a utility's costs can realistically be assigned to one particular function. Many of the costs are joint costs that are due jointly to two or more functions. For example, the cost of maintaining a pumping station cannot be held due to one particular measure of demand, be it year-round average demand, peak-day demand, or peak-hour demand; rather, this cost depends upon all three of these demands and cannot be allocated to any one function in a way that has any economic significance. The fault of both methods is that, quite groundlessly, they seek to allocate joint costs among separate functions. [10]

The next step in the procedure is to divide the costs allocated to each function by the quantity of water handled by that function to give the per unit cost of using that function. In the Commodity/Demand method the cost of providing a unit of water is found by dividing the total costs allocated to Commodity Costs by the total quantity of water delivered by the utility.

Similarly, the cost of providing a unit of capacity to meet the peak-day demand is found by dividing the Peak-day Demand Costs by the total quantity of water demanded by the system over the peak day; the cost of each unit of peak-hour capacity is the Peak-hour Demand Cost divided by the total water quantity demanded during the peak hour. In the Base/Extra Capacity method, the cost of providing for each unit of year-round average demand is the total Base Capacity Cost divided by the year-round average quantity demanded; the cost of providing each unit of extra capacity to meet the peak-day demand is the Peak-day Extra Capacity Cost divided by the difference between the total quantity demanded by the system on the peak day and the year-round average demand; the unit cost of the peak-hour extra capacity is the Peak-hour Extra Capacity Cost divided by the difference between the total quantity demanded during the peak hour and the year-round average demand. In both methods, the cost of serving each customer is found by dividing the total Customer Costs by the total number of customers served.

Although these costs are expressed as a cost per unit of water delivered, a cost per unit of peak capacity, and so on, they do not indicate the costs resulting from a unit increase in total water consumption or a unit increase in the peak demand on the system. These costs are merely an average of the allocated costs and have no economic significance. Schedules of water rates based on these costs will not charge customers for water in a way that reflects the costs that these customers impose on the system.

The real economic cost of a unit increase in total water consumption, or a unit increase in the peak-day or peak-hour demand, should be estimated by determining how the utility's total costs would be changed by a unit change in each measure of water demand, the other measures remaining constant. These costs are the marginal costs of each measure of output. The marginal commodity cost MCQ is the increase in the utility's total costs resulting from a unit increase in the amount of water delivered when the peak demands and the number of customers are held constant. Let $\triangle$ signify a change, C equal total cost, and Q equal water quantity.

Then,

$$MCQ = \triangle C/\triangle Q \qquad (4)$$

The marginal peak-day cost is the increase in the utility's total costs resulting from a unit increase in the peak-day quantity demanded when the amount of water delivered, the peak-hour quantity demanded, and the number of customers are held constant; that is

$$MCD = \triangle C/\triangle Q_D \qquad (5)$$

The marginal peak-hour cost,

$$MCH = \triangle C/\triangle Q_H, \qquad (6)$$

is defined in a similar way, and the marginal customer cost MCN is the increase in the utility's total costs resulting from changing the number of customers connected to the system when the demand for water remains constant; that is

$$MCN = \triangle C/\triangle N \qquad (7)$$

These marginal costs could be determined by examining the way in which the cost of financing new bonds and operating the various elements of a water system varies when the amount of water delivered, the peak-day and peak-hour demands, and the number of customers are each increased by a unit amount, the other measures remaining constant. The marginal commodity cost is the cost of obtaining an additional unit of water, plus the capital and operating costs of treating and delivering this water if the capacity of the system to deliver peak demands is not affected. Depending on the particular circumstances of the utility, the cost of obtaining additional water may be the cost of increasing the capacity of the utility's reservoir system to yield an additional unit of water, the addition of a groundwater well, or simply the cost of purchasing additional water from another agency. The marginal peak-day capacity cost is the cost of financing and maintaining the additional transmission mains, pumping plant, and treatment facilities needed to meet an increase in the peak-day demand. Similarly, the marginal peak-hour capacity cost is the cost of financing and maintaining facilities such as distribution mains, service reservoirs, and pumping plants needed to meet an increase in the peak-hour demand on the system. The marginal customer cost is the cost of providing additional meters and connections and of administering an additional account. These marginal costs could be used as a basis of rate schedules that charge customers for water in a manner that reflects the cost of providing them with service.

A final problem with the AWWA's allocation methods is that they are used as a basis for a further allocation of costs, this time among the utility's various classes of customers. Generally speaking, commercial customers use water much more uniformly than do residential customers. Accordingly, most utilities have separate rate schedules for residential and commercial customers, recognizing the differences in costs of serving these two classes of customers. In many cases, customers with a particularly uniform pattern of consumption are designated as industrial customers and are charged according to another, industrial, rate schedule.

The costs allocated to each function are allocated among the three classes of customers in proportion to the extent to which they are assumed to contribute to each measure of quantity demanded. In the Commodity/Demand method, Commodity Costs are allocated among the customer classes in proportion to total water consumption by each class; the Peak-hour Demand Costs and the Peak-day Demand Costs are allocated in proportion to the extent to which use by each customer class contributes to the peak-day and peak-hour demand on the system. In the Base/Extra Capacity method, the Base Capacity Costs are allocated in proportion to total water consumption by each class, and the Peak-day and Peak-hour Extra Capacity Costs are allocated in proportion to the extent to which use by each customer class is assumed to contribute to the excess of peak-day and peak-hour demand over the year-round average demand.

The purpose of this allocation of costs is to determine the costs of serving each class of customer. These costs are compared with revenues that the final rate schedules are expected to raise from each class of customer as a. ssurance that these rates are equitable. However, a utility's costs can no more be allocated among classes of customers than can these costs be allocated among functions. In addition, since the contribution to peak-day and peak-hour demand is simply assumed because no measurements are available, the allocations become arbitrary allocations of arbitrary costs.

The Design of Water Rate Schedules

The final step in the AWWA's procedure is to use the unit costs determined in the first part of the procedure to design the actual rate schedules. The objective is to design a multiple-part tariff that recovers costs from customers in accordance with the way that their use of water contributes to the required capacity of the system. The tariff is designed by first assuming how the ratios of a class of customers' peak-day and peak-hour demands vary with their monthly consumption. This assumed measure of customers' peaking behavior is used to establish ranges of consumption forming the use blocks of the tariff and to set "peaking ratios" for each use block. For example, the utility staff might assume that customers using fewer than 5,000 gallons per month have a peak-day demand of three times their average daily consumption and a peak-hour demand of four times their average consumption, and that customers using between 5,000 gallons per month and 10,000 gallons per month have a peak-day demand of twice their average daily consumption and a peak-hour demand of three times their average daily consumption. These peaking ratios are then used in combination with the unit commodity costs and the unit peak-day and peak-hour capacity costs to set the prices to apply in each use block of the tariff. Traditionally, this step in the procedure has given a decreasing multiple-part tariff, for it has been assumed that a customer's peaking ratios fall as that customer's consumption increases. The assumptions of Carollo Engineers, as shown in table 4-1 for Tucson customers, were just the opposite.

This method of designing rate schedules is unsatisfactory in two respects. First, as already discussed, the costs that are used as a basis for the prices in the tariffs have little economic significance, for they are unit-allocated costs, not marginal costs. Second, the way that the costs are used to set the prices in the tariffs are not closely related to the way that a customer's use of water imposes costs on the utility. A water system is designed to meet the peak demands occurring during the summer months. When any customer uses water during the peak day or the peak hour, the peak demand on the system will be increased, and in the long run, additional capacity must be provided. Any customer's use of water during the summer will impose capacity costs on the system, since this use presses against the capacity of the system, creating a need for additional capacity. In contrast, a customer's use of water during the winter will have no effect on the capacity that must be provided because the system is designed to meet the higher levels of consumption occurring during the summer. Thus, winter use does not create a need for additional capacity and so does not impose capacity costs on the system. However, in the AWWA's procedure, the unit capacity costs are used to set prices in a tariff that is in force all year round. Thus, costs resulting solely from use of water during the summer are recovered partly from bills for winter consumption. This procedure tends to underprice water used during the summer and thus to encourage high summer water use.

A number of water utilities recently have introduced innovative rate schedules. The most popular forms of innovation are increasing multiple-part tariffs and seasonal rates that charge customers extra for summer consumption that exceeds a specified proportion of their average winter consumption. Regrettably, these rate schedules are designed by applying the traditional procedures in a novel manner and so are subject to virtually all of the deficiencies inherent in the traditional approach. The increasing multiple-part tariffs are based on the assumption that a customer's peaking ratios increase as that customer's average consumption increases. Increasing multiple-part tariffs are a partial solution to the underpricing of summer water inherent in tariffs that are in force all year round; since a customer's consumption is generally higher in the summer than in the winter, the customer will be in a higher use block and so will face a higher price in the peak summer months.

Generally speaking, however, the thinking behind these innovative seasonal rates does not embody a sound understanding of how the seasonal variation in water consumption imposes costs on a water system. The rates seem to be based on the belief that the cost of providing for the extra capacity needed to meet the peak demand on the system should be borne solely by those customers who increase their use of water during the peak. However, *all* customers who use water at the time of the peak contribute to the peak, regardless of how much water they may use at some other time of the day or year. Sometimes it is thought that the cost of providing for the peak should be recovered by charging extra for consumption in excess of a customer's winter consumption because this use "causes" the peak. Again, all use of water during the peak contributes to the peak, regardless of whether this use does or does not exceed a customer's winter consumption. In any case, the costs used to set the tariffs are unit-allocated costs that do not represent the real costs imposed on the system by a customer's use of water, and thus do not charge customers for water in an equitable and efficient manner.

The following section describes how a utility could design a water rate schedule that would charge customers so as to reflect the costs that these customers impose on the water system. We then compare this optimal type of schedule with the actual rate schedules used by the City of Tucson.

Marginal-cost Pricing

A customer imposes cost on a water system in three ways. First, the customer's use of water means that the utility must deliver an additional amount of water. Second, any use of water by the customer during the time of the peak-hour or peak-day demands on the system contributes to these peak demands, meaning that additional capacity must be provided. Third, the presence of the customer on the system means that an additional connection and meter must be provided and that an additional account must be administered. These are the utility's marginal costs, as defined earlier in this chapter. They could be used to set schedules of water rates that depend on the way in which a customer's use of water contributes to the peak demand on the system.

A customer's contribution to the peak demands are that customer's consumption over the peak day of the year and his or her consumption during the peak hour of the year. If it were practicable to measure these peak-time demands directly, by means of time-of-day meters, the utility would be able to establish a rate schedule that would charge customers directly for their individual contributions to the peak demands on the system. A rate schedule of this kind would consist of two components: the first would be a monthly two-part tariff comprising a charge for the provision of service composed of the marginal commodity cost MCQ and the marginal customer cost MCN; the second would be an annual capacity assessment consisting of a charge equal to the marginal peak-day capacity cost MCD for each unit of water used over the peak day, plus a charge equal to the peak-hour capacity charge MCH for each unit of water used during the peak hour. MCN and MCQ are the *monthly* costs of serving an additional customer, while MCD and MCH are the *annual* costs of providing additional units of capacity. A customer's monthly bill would be

$$B = MCN + MCQ(Q) \qquad (8)$$

where Q is the customer's monthly consumption. The customer's annual capacity assessment would be

$$A = MCD(Q_D) + MCH(Q_H) \qquad (9)$$

where Q_D and Q_H are the customer's consumptions over the peak day and during the peak hour, respectively.

In most cases it is not practicable to measure a customer's peak-time demands directly; only the customer's total consumption each billing period is measured. However, customers could be charged indirectly for their contribution to the peak demand on the system if some assumptions were made regarding customers' peaking patterns and these assumptions were used to design a tariff to be applied to customers' total consumption each billing period. The AWWA's procedure is to assume that the ratios between a customer's peak-time consumption and that customer's total consumption have some

particular values within a given range of consumption. This approach is valid, provided the rate schedule is designed with a clear understanding of how a customer's peak-time consumption contributes to the peak demand on the system and imposes costs on the utility.

Suppose a utility found that customers using between 5,000 and 10,000 gallons per month generally have a rate of consumption over the peak day of the month of twice their average daily consumption, and a rate of consumption during the peak hour of that month of three times their average daily consumption.[11] If it is assumed that the ratios between the customer's peak-time demands and his or her average daily consumption are fixed over this range of consumption, it follows that an increase of 1,000 gallons in the customer's total consumption in a given month (that is, an increase of 33.3 gallons per day in the customer's average daily consumption that month) will result in an increase of 2(1,000/30) gallons per day in the peak-day demand on the system, and an increase of 3(1,000/30) gallons per day in the peak-hour demand on the system.

The costs that this increase in consumption imposes on the system depend upon whether the month in question is the peak month of the year. An increase in the peak-day and peak-hour demands during the peak month of the year will mean that additional capacity must be provided to meet this demand. However, an increase in the peak demand during any other month of the year will make no difference to the required capacity of the system, since capacity is already provided to meet demand during the peak month.

Consequently, an increase of 1,000 gallons in the customer's consumption during the peak month imposes a cost on the system equal to the cost of delivering an additional 1,000 gallons, plus the cost of providing an additional 2(1,000/30) gallons per day of peak-day capacity, plus the cost of providing an additional 3(1,000/30) gallons per day of peak-hour capacity. During any other month of the year, the cost imposed by a 1,000-gallon increase in the customer's consumption is merely the cost of delivering an additional 1,000 gallons.

These costs could be used to design a rate schedule reflecting the variation in water consumption over the course of the year. If, for example, it were known that the peak demands on the system always occurred in June, the tariff in force during June should be set to reflect delivery and capacity costs, while the tariff in force during the other eleven months of the year should be set to reflect only delivery costs. For example, the price of a unit of water for a customer using between 5,000 and 10,000 gallons per month during the month of June would be

$$P = MCQ + (2/30)MCD + (3/30)MCH \quad (10)$$

The prices in the other blocks of the June tariff would be set in a similar manner, using the peak ratios for these use blocks. This procedure will give a multiple-part tariff that may be increasing or decreasing, depending upon whether the customer's peaking ratios increase or decrease as his or her monthly consumption rises. The tariff would also include a fixed charge for the provision of service equal to the marginal customer cost MCN. The tariff for the remaining eleven months of the year would be simply a two-part tariff comprising a uniform price equal to the marginal commodity cost MCQ and a fixed charge equal to the marginal customer cost MCN.

In most actual circumstances, peak demand on the system does not always occur during the same month of the year, but rather may occur at any time during the summer months. Thus, the year could be divided into two seasons: a peak summer season consisting of the two or three months during which the peak demand on the system may occur, and an off-peak winter season consisting of the remaining months of the year. The summer tariff could be a weighted average of the peak-month and off-peak-month tariffs. The weights would be in proportion to the probabilities with which the peak demand occurs. If it is assumed that the peak demand occurs with equal probability in each of the three months of the summer season, this monthly cost would be

$$\begin{aligned} C &= \tfrac{1}{3}\,[MCQ + (2/30)MCD \\ &\quad + (3/30)MCH] + \tfrac{2}{3}(MCQ) \quad (11) \\ &= MCQ + 2MCD/(3 \times 30) \\ &\quad + 3MCH/(3 \times 30) \end{aligned}$$

Table 4-2 presents an example of how a seasonal rate schedule consisting of a summer multiple-part tariff and a winter two-part tariff could be developed from a utility's marginal costs and information on customer's peaking patterns. The thinking behind the summer tariff is quite different from that behind the increasing multiple-part tariffs that have been adopted in recent years, including the rate schedules for Tucson. Those tariffs appear to be based on a belief that the peak demand on the system is "caused" by the customers who increase their use of water during the summer. This belief is erroneous, since all customers using water at the time of the peak contribute to the peak. In contrast, our proposed summer tariff is an indirect form of peak-load pricing based on the assumption of a fixed relationship between a customer's total monthly consumption and that customer's peak-time demand. A customer using a large amount of water is charged more for a unit of water than a small customer is, because the assumed peaking ratios indicate that the large customer's contribution to the peak demand on the system will increase more when that customer uses an additional unit of water than will a small customer's contribution. For example, when a customer in the 20,001- to 50,000-gallon-per-month use block uses an additional 1,000 gallons, the peak-day demand on the system will increase by 4.8(1,000/30) gallons per day. However, when a customer in the 5,001- to 10,000-gallon-per-month use blocks uses an additional 1,000 gallons, the peak-day demand is increased by only 2.0(1,000/30) gallons per day.

An increasing tariff of this kind should be used only in cases showing definite evidence that consumption by large customers is relatively more strongly peaked than is consumption by small customers. In cases lacking evidence that customers' peaking ratios vary with consumption, the summer tariff should be a two-part tariff with a uniform price based on the peaking ratios assumed to apply to all levels of consumption.

The rates developed in table 4-2 are a departure from traditional practice in that the summer

Table 4-2. Design of Rate Schedules Using Marginal Costs

1. Marginal costs (hypothetical)

Cost	$	Description
Marginal commodity cost MCQ	0.15	Cost of delivering an additional 1,000 gal
Marginal peak-day capacity cost MCD	20.00	Annual cost of providing an additional 1,000 gal/day of capacity
Marginal peak-hour capacity cost MCH	30.00	Annual cost of providing an additional 1,000 gal/day of capacity[a]
Marginal customer cost MCN	2.00	Monthly cost of providing for an additional customer

2. Summer tariff

Use block (gal/month)	Peaking ratios		Price per 1,000 gal			
	Peak day K_D	Peak hour K_H	Commodity component MCQ	Peak-day components $K_D MCD/(3 \times 30)$	Peak-hour component $K_H MCH/(3 \times 30)$	Total
0 to 5,000	1.8	2.6	0.15	0.40	0.87	1.42
5,001 to 10,000	2.0	3.0	0.15	0.44	1.00	1.59
10,001 to 20,000	2.4	3.8	0.15	0.53	1.27	1.95
20,001 to 50,000	3.0	4.8	0.15	0.66	1.60	2.41
More than 50,000	3.4	5.6	0.15	0.76	1.87	2.78

Notes: The winter tariff should consist of a charge of $0.15/1,000 gal for all water delivered, plus a charge of $2.00/month for the provision of service.

The summer tariff is for a three-month session. The general formula for the price in the use block of a summer tariff is $P = MCN + K_D MCD/(n \times 30) + K_H MCH/(n \times 30)$, where K_D and K_H are the peaking ratios for the use block in question and n is the number of months in the season in which the peak demands may occur. The summer tariff should also include a charge of $2.00/month for the provision of service.

In many cases, different classes of customers have different peaking patterns; for example, consumption by commercial customers is usually more strongly peaked than is consumption by residential customers. In these circumstances, peaking ratios for each class of customer should be estimated and used to design a summer tariff for each class of customer.

[a]Marginal peak-hour capacity is measured in terms of additional rated capacity per *day*.

tariff is much higher than the winter tariff. This situation arises because use of water during the summer imposes high capital costs on a water system, while winter use imposes only the relatively low operating costs. A comparison of these rates with the rate schedules commonly in use clearly shows how the traditional approach to rate making underprices water used during the high-use summer months.

Revenue Requirements

When a utility's rate schedules are based on the utility's total average cost per unit of water delivered, it is guaranteed that the revenues raised by the rates will meet the utility's revenue requirements. When the rate schedules are based on service-specific marginal costs, and if net economies of scale exist, total revenue requirements may not be met. In today's developed economy where water itself is becoming more expensive to obtain, in most cases the sum of a utility's marginal costs will exceed its average cost, meaning that rates based on marginal costs will raise more revenue than is required to meet current requirements. In these circumstances the rate schedule could be altered to make the revenues raised by the adjusted schedule equal to the utility's revenue requirements. Any adjustments should be carried out with the objective of affecting water consumption as little as possible. Since the fixed service charge does not affect the amount of water purchased by a customer, a convenient way of adjusting the rates is to alter the service charge. If the excess revenues are too great to be nullified by eliminating the service charge, the price for the first few units of water used each month could be reduced to a nominal or zero price.[12] An adjustment of this kind will have little effect on water consumption, since most customers would face the unadjusted price prevailing at higher levels of consumption. In this way, systematic modifications to the rate schedules can be used to reconcile differences between a utility's revenue requirements and revenues raised by rates based on the utility's marginal costs.

However, if there is a reason for conserving water for the future, that is, if a nonrenewable resource is being depleted as in the case of Tucson, surplus revenue should not be regarded as a problem. The object should be to price water so that water service is not priced lower than its real cost. If excess revenues are obtained, they could obviously be used to provide other social services, such as compulsory education, for which society has decided that the social benefits obviously exceed the private costs. Providing subsidized education is a question much different from that of providing subsidized water. Almost all society benefits from an educated, civilized society. Only some people benefit from more water development than society would be willing to pay for on an individual private basis.

Tucson Utility Rate Schedules

We have shown that the real costs of water service depend on the cost of delivery of an additional unit of water, the costs of providing additional units of peak-day and peak-hour capacity, and the cost of providing for an additional customer.

Since there is never any capacity problem in the winter months, the only costs are the water itself and the cost of providing service. Since each unit of water contributes equally to the additional water cost, economic reasoning suggests a single rate for each unit of winter water use. There is no rationale for an increasing block rate schedule in the winter other than a general desire to hold down total water use regardless of its costs or benefits. This desire would reflect the preservationist point of view.

Peak-load costs occur only in one or two summer months. Even then, since all customers using water at the time of the peaks contribute to the peaks, an increasing block rate schedule will not reflect the real costs of delivering water unless the larger water users' peaking ratios (the

ratio between peak use and average use) are higher than those for a smaller water user. Only in this case will the larger water users' contribution to peak demand on the system increase more when an additional unit of water is used than will the small water users' contribution. The evidence in Tucson indicates that the peaking ratios are constant between users of small and large quantities of water. Thus, a marginal-cost pricing procedure suggests a high flat rate in the peak summer months reflecting marginal peaking costs, and a lower flat rate during the rest of the year simply reflecting the cost of water and customer service.

Tucson water utility rate schedules from 1925 through 1982 are summarized in figures 4-2 and 4-3. Until 1964, the City of Tucson Water Department delivered water at a flat rate, summer and winter, after charging a minimum fee for the first 700 ft^3. In 1964, a two-step declining rate schedule was introduced, again with the minimum charge for the first 700 ft^3. Declining-step schedules were in force for the next ten years.[13]

The first increasing-step tariff was introduced in 1974. A minimum was maintained, but only for 600 ft^3. After the minimum there were two steps that applied in both summer and winter. In defense of this schedule, the second step began at 1,400 ft^3, about mean yearly water usage for single-family dwellings; therefore, most families would be in the second step in the summer and the first step in the winter.

Until 1976, the rate schedules were designed simply to cover revenue needs. Water conservation was not a goal of the Tucson Water Department. However, the two-step increasing schedule was an attempt to recognize that extra capacity was needed in the summer. John Carollo Engineers had designed the new schedule using the American Water Works Association's Base/Extra Capacity method, described in the first part of this chapter.

A many-step increasing rate schedule was the product of the political events of July 1976, described in chapter 2. The minimum water allowance was dropped, although a monthly service charge was introduced. In addition, a pumping service charge was introduced so that

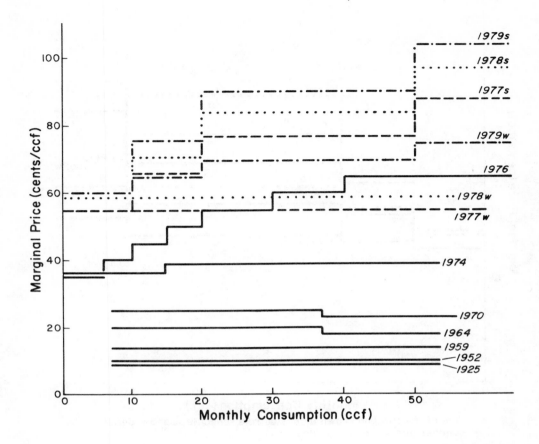

Figure 4-2. Water rates for single-family residences in Tucson, 1925–1979. (ccf = 100 ft^3)

Note: Year of introduction shown alongside rate schedule; s and w denote summer and winter schedules.

water was delivered to areas at higher elevations only at higher cost. Despite the political furor that this new rate schedule engendered, the multiple-part increasing tariff (but without the pumpage lift surcharge) has been the basic model since that time. In 1977 and 1978 Tucson returned to a single rate for the winter months, but with increasing steps in the summer. Since April 1979 both summer and winter rates have been multiple-part increasing tariffs, although winter rates are at a lower level. The Base/Extra Capacity rate-making system is still in use. The inverted block structure is now considered necessary to encourage year-round water conservation. Summer is de-

fined as May through October, winter as November through April.

Tucson's Rate Schedule Versus Marginal-cost Pricing

For a rate schedule to be economically efficient, each individual should pay neither more nor less than the marginal cost of providing water services. Under these circumstances, neither too little nor too much water service will be provided. Further, at least in one sense of the word equity, no consumer should subsidize any other. We demonstrated in the first part of this chapter

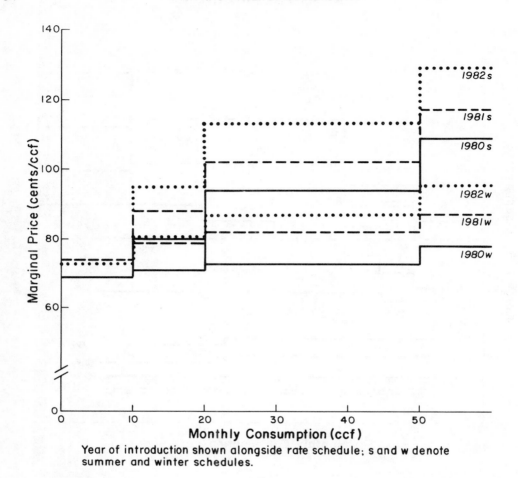

Figure 4-3. Water rates for single-family residences in Tucson, 1980–1982. (ccf = 100 ft³)

how an efficient and equitable schedule could be constructed. What is the situation in Tucson?

Tucson water rates are set using a form of the Base/Extra Capacity method suggested by AWWA. This method relies on an arbitrary allocation of historical average costs and thus cannot be efficient. Since it is almost surely the case that current marginal costs of water in Tucson are higher than total average cost, this procedure encourages water use and extra development for water use, a peculiar situation for an area that now preaches the need for water conservation.

Both winter and summer schedules are increasing-block schedules. There is no rationale for an increasing-block winter schedule. This

schedule should be simply a two-part tariff comprised of a uniform price for all, equal to the marginal commodity cost MCQ, and a fixed charge equal to the marginal consumer cost MCN. The increasing-step function in the nonpeak winter months cannot be economically efficient, and asks the large users to subsidize the smaller users. Further, since the daily peak always occurs in June or July, summer rates are in force for too long and winter rates for too short a time period.

An increasing-block summer schedule *could* be economically efficient and equitable if the ratios of peak-day and peak-hour demands were positively associated with total water consumption, and if the schedule were carefully constructed using that knowledge. Carollo assumed a positive association in his 1976 report to the Metropolitan Utilities Management Agency, but hard evidence was lacking and the schedules are not so constructed.[14]

In fact, using monthly water use data from a sample of individual single-family residences, we found that the ratio between water demand in the peak month of the year and mean monthly water use is constant over all use blocks (table 4-3 and figure 4-4).[15] Regression analysis showed that from 62 to 76 percent in the variance of individuals' peak monthly use could be explained by a simple linear relationship between peak monthly use and mean annual monthly water use (see R^2 column in table 4-3). Use of a curvilinear function did not improve the explanation of peak use as approximately twice mean monthly use for all levels of water consumption.

It is possible that the ratios of peak-day and peak-hour use could be positively associated with total water consumption even if the ratio of peak-month use to total consumption is constant. However, we believe such a situation unlikely. Thus, Tucson's increasing-block schedule probably is neither efficient nor equitable. It will tend to hold down water use, and that is the reason stated by the utility for its adoption.[16] On the other hand, since winter rates are also step functions and are not greatly lower than

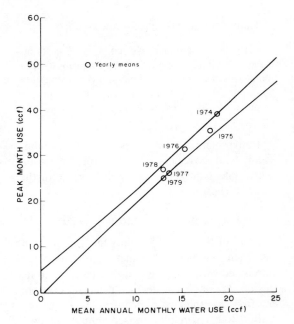

Figure 4-4. Peak-month use as a function of mean annual monthly use, 1974 to 1979. (All six years are linear functions within the limits of the two lines shown.) (ccf = 100 ft³)

summer rates are, winter use subsidizes summer use and tends to increase summer consumption above what it would be under marginal-cost pricing.

Table 4-3. Peak-month Use As a Function of Mean Annual Monthly Use, 1974 to 1979

Year	Intercept[a]	Slope coefficient[a]	R^2	Number of observations	Mean of maximum water use	Mean of average water use	Maximum observed water use	Water use for 90 percent of sample	Peak month as a percent of mean month
					(———————— 100 ft³/month ————————)				
1974	2.208 (0.587)	1.979 (0.027)	.72	2,121	39.147	18.663	510	<68	210
1975	4.738 (0.442)	1.711 (0.021)	.76	2,117	35.437	17.937	385	<61	198
1976	3.327 (0.526)	1.826 (0.025)	.71	2,127	31.526	15.444	215	<56	204
1977	−0.321 (0.526)	1.959 (0.034)	.62	2,132	26.169	13.519	474	<45	193
1978	0.554 (0.451)	2.010 (0.029)	.69	2,147	27.660	13.484	275	<48	205
1979	1.690 (0.351)	1.811 (0.023)	.75	2,145	25.794	13.308	219	<46	193

Note: Monthly water use data from random sample of single-family residences, Tucson Water Department records.
[a]Standard errors are shown in parentheses below the estimated coefficients.

Hanke comments that such a system

is inefficient, discriminatory, and perverse in its income-redistributive properties. Although most environmentalists and those in the industry view increasing-block rates and block seasonal rates as an economist's dream come true, the economist views them as a nightmare.[17]

Hanke blames economists for not communicating with those persons responsible for setting water rate policy. He is surely correct in the case of Tucson.

The Cost of "Scarce" Water

In all of the discussion thus far, implicitly it has been assumed either that water was being purchased by a utility for further delivery to its customers, or that water itself had no cost. In the first case the cost of acquiring water by the utility simply would be passed along to its customers as a marginal commodity cost MCQ. In the second case there would be no marginal commodity cost associated with the water itself, but only with the cost of pumping that water, or transporting it, or both, from its source to the consumer.

It is curious that in Tucson, where water presumably is in scarce supply (at least in the long run), water itself has no cost—the only costs to consumers are the costs of pumping the water to the surface from the underground aquifer and of delivering that water to the users. The effect of such a pricing system is to encourage water use today by current residents, to hurry new use through population growth, and essentially to assume that water has no future value. When the marginal cost of any resource is zero, people consume that resource until they are physically satisfied. In Tucson, people's water use may be affected by the cost of acquisition and distribution systems, but not by the cost of water itself. This pricing policy is completely in accordance with the traditional goal of water utilities to fulfill their customers' physical "needs" or "requirements" without reference to conservation because of resource scarcity.

Because groundwater in place has always been a common property resource in Arizona, it has never had a price. People could pump as much

water as they wished from beneath the surface of their property for "reasonable" use on their overlying lands. They could not, however, sell that water to be transferred elsewhere for some other use. Since passage of the Groundwater Transfer Act of 1977, under special, regulated circumstances groundwater may be transferred away from the land from which it was drawn, but the water itself still is not sold. The right to transfer water will be granted only if the farmland previously using the water was bought or leased, and retired from agriculture. It is the land that is being purchased, at the land's opportunity cost in agriculture.

Passage of the Arizona Groundwater Management Act of 1980 and the creation of a Department of Water Resources may eventually lead to a changed view of pricing water. The law mandates measuring devices even on agricultural wells, and states that measures must be taken to achieve a safe yield in the groundwater aquifer surrounding Tucson early in the next century. (Safe yield is defined as achieving a balance between average annual groundwater recharge and average annual pumpage.) Under current conditions groundwater may be viewed as scarce by the public, but in fact there is no limit to how much may be pumped, and thus the water is free in the economic sense of the word. It might become limited in physical quantity in the future, but that future is far away and not defined. If safe yield, by political fiat, must be achieved, the quantity of pumped water that may be used becomes a defined number. The right to pump that water should become very valuable, and the water itself should acquire a price.

What price might the citizens of Tucson currently assign to water itself if they were truly interested in water conservation and wished to incorporate that price into the marginal commodity cost MCQ of the rate schedule? Surely the groundwater should be priced at least as high as what the city is willing to pay to achieve an additional supply of equal quality. Tucson has applied for an allocation of surface water to be delivered through the Central Arizona Project (CAP) currently under construction. Payment contracts have not been signed, nor have estimates been made of the costs of constructing water quality improvement facilities and other facilities necessary for receiving the water, for which Tucson must pay. Thus, the exact marginal commodity cost of water is not known. However, one may think in orders of magnitude. It has been suggested that the raw CAP water might be delivered to Tucson at a price of around $100 per acre-foot. This price translates to $0.23 per 100 ft^3. Other commodity costs (as distinguished from peaking costs) of accepting this marginal water surely will at least double the basic $0.23 per 100 ft^3. Stephen Davis of the Tucson Water Department suggests a total cost of about $250 per acre-foot or $0.58 per 100 ft^3.[18]

The current (August 1983) Tucson water rate schedule has a summer water price of about $1.00 per 100 ft^3 in the mean use block for single-family residences. Assuming that the $1.00 cost is based on marginal costs rather than being an average cost (it is not, and thus is too low), an additional charge of at least $0.58 per 100 ft^3—an increase of at least 58 percent—should be added.

If the rate schedules were constructed on the system of marginal-cost pricing, an immediate price increase of that general magnitude would be both economically efficient and equitable. It would be a conservation program in the rational sense of not developing or using water for which people were unwilling to pay. It would conserve water for the future but allow development as people were willing to pay for it.

The current pricing schedules of the City of Tucson Water Department are neither efficient nor equitable, and they encourage rapid water development through limited attempts to encourage current conservation practices.

Notes

1. Charles W. Howe and F. P. Linaweaver, Jr., "The Impact of Price on Residential Water Demand and Its Relation to System Design and Price Structure," *Water Resources Research* vol. 3, no. 1 (1967) pp. 13–32. The transmission plant brings the water from its source to the distribution area; the distribution plant distributes the water to the individual customers.

2. Steve H. Hanke, "Pricing Urban Water," in S. J. Mushkin, ed., *Public Prices for Public Products* (Washington, D.C., Urban Institute, 1972) pp. 283–306.

3. Russell F. Scareto, "Time-Capacity Expansion of Urban Water Systems," *Water Resources Research* vol. 5, no. 5 (October 1969) pp. 929–936.

4. Howe and Linaweaver, "The Impact of Price."

5. John Carollo Engineers, "Report Submitted to the Metropolitan Utilities Management Agency," Final report (Tucson, Ariz., March 11, 1976); J. E. Flack and G. J. Roussos, "Water Consumption Under Peak-Responsibility Pricing," *Journal of the American Water Works Association* vol. 70 (March 1978) pp. 121–126; Fred P. Griffith, Jr., "An Innovative Approach to Rate-Making," *Journal of the American Water Works Association* vol. 69 (February 1977) pp. 89–91.

6. Carollo Engineers, "Report to MUM"; Flack and Roussos, "Water Consumption."

7. Steve H. Hanke, "Water Rates: An Assessment of Current Issues," *Journal of the American Water Works Association* vol. 67 (May 1975) pp. 215–219.

8. Ibid.

9. American Water Works Association, *Water Rates* AWWA Manual M1 (2nd ed., Denver, Colo., 1972).

10. One must distinguish between the average historical cost of delivering a unit of water—a perfectly legitimate economic concept—and the "average" per unit cost of a particular function such as providing peak-day capacity. The latter concept is arbitrary and has no economic significance. An analogy may help explain the difference. If a farmer produces only apples, the average cost of producing a bushel of apples can be computed. However, if a farmer uses the same machinery to produce apples, oranges, and grapefruit, the total cost of production can be computed, but the average cost of production of each product cannot be computed correctly, since that cost would depend on the arbitrary selection of a method of dividing up the capital machinery costs.

11. Peak-hour capacity conventionally is rated in gallons per day, not in terms of actual flow or pumpage. It is the capacity of the system to deliver, at any hour, a given percentage of the average daily consumption.

12. A negative service charge is another reasonable alternative.

13. The rate schedules are expressed in 100 ft^3. One cubic foot equals 7.38 gallons. One acre-foot equals 325,851 gallons or 43,600 ft^3.

14. Carollo Engineers, "Report to MUM."

15. The sample is described in chapter 5.

16. Interview with Stephen Davis, Tucson Water Department, Tucson, Ariz., February 3, 1982.

17. Steve H. Hanke, "Pricing as a Conservation Tool: An Economist's Dream Come True?" in David Holtz and Scott Sebastian, eds., *Municipal Water Systems* (Bloomington, Indiana University Press, 1978) p. 238.

18. Interview with Stephen Davis, Tucson Water Department, Tucson, Ariz., February 10, 1982.

5

Price and Consumer Demand

Economic Demand and Water Use

Recent records of the City of Tucson Water Department show that since fiscal year 1965/66, per capita water consumption in Tucson rose steadily until 1969/70, fluctuated up and down in the 1971/72 to 1975/76 period, and then dropped continuously from 1976/77 through 1978/79 (table 5-1, col. 5). In 1978/79 per capita water consumption reached a low of 147.5 gallons per day.[1]

Reduction in per capita use was so dramatic that total average daily pumpage by the water utility declined from a high of 75.4 million gallons per day in fiscal year 1973/74 to a low of 59.4 million gallons per day in 1977/78 despite a rapidly growing service population (table 5-1, cols. 2 and 4). Total pumpage declined by 21.2 percent during this period, while the service population increased by 13.3 percent.

What has not been highly publicized, however, is that in each of the last three fiscal years, 1979/80 to 1981/82, per capita water use has again risen (table 5-1). The combination of rising per capita use along with a rising service population has caused total pumpage to increase

to the levels immediately preceding 1976/77 levels.

The political events of the summer of 1976, followed by continued publicity about water problems, have been cited as the major force in reducing per capita consumption of water. But as the following analysis shows, many of the changes in per capita water use—both decreases and increases—can be attributed directly to price itself. Price *is* an effective regulator of people's water usage, especially when knowledge about water price is reinforced by preachments and political events. Thus, the institution of a rational marginal-cost pricing scheme, as discussed in chapter 4, could play an important role in the water conservation programs of Tucson and other municipalities.

We attempt to estimate the effects of changing water prices by estimating Tucson residents' demand curve for water. A demand curve is the relationship between the quantities that would be purchased at alternative prices. If water is like almost every other commodity in the world, less will be purchased at higher prices, and vice versa. A static demand curve gives this relationship at a particular point in time, given the

Table 5-1. City of Tucson Water Department's Pumpage, Fiscal Years 1965/66 to 1981/82

Fiscal year	Average daily pumpage (million gal/day)	Number of active services	Service population[a]	Pumpage per capita (gal/day)
1965/66	41.5	60,934	232,000	179.2
1966/67	41.8	62,383	237,000	176.3
1967/68	43.8	63,483	241,000	181.6
1968/69	48.3	68,063	259,000	186.7
1969/70	51.2	71,241	264,000	194.2
1970/71	54.1	78,177	289,000	187.0
1971/72	57.8	84,816	314,000	184.2
1972/73	60.1	95,105	352,000	170.8
1973/74	75.4	99,604	369,000	204.6
1974/75	67.6	102,813	380,000	177.7
1975/76	70.0	105,595	370,000	189.4
1976/77	60.5	109,480	383,000	157.9
1977/78	59.4	113,105	396,000	150.1
1978/79	60.8	117,777	412,000	147.5
1979/80	66.2	122,514	429,000	154.4
1980/81	69.7	125,367	439,000	158.9
1981/82	74.0	127,650	447,000	165.6

Source: City of Tucson Water Department, 1982.

[a]Assumes 3.7 people per service through 1974/75; 3.5 people per service are assumed thereafter.

surrounding conditions of incomes, the prices of other goods and services, and the tastes and preferences of consumers. A price change will cause a change in the *quantity demanded*. A change in any of the surrounding conditions, prices remaining constant, will cause a *shift in demand*; that is, the demand curve (demand) shifts to the left or right.

We attempt to separate the effects of price versus those of preachments and politics by estimating the static demand curve and then testing how well this curve explains consumer behavior. We find that price alone explains behavior in some years but not in others. In certain years the whole demand curve has apparently shifted to the left. This shift in demand means that conservation behavior in Tucson is more than a simple reflection of higher prices. People in Tucson, where demand has shifted, have altered their tastes and preferences and are evaluating water in a new way. Presumably their tastes and preferences for water were altered because of the publicity surrounding the political events beginning in 1976 and the continuing campaign by the Tucson Water Department to "Beat the Peak," that is, to lower water use during the peak-use times of the day.

Demand Estimation

Our general hypothesis is that the quantity of residential water that will be demanded by an individual family is a function of the price of the water, the family's income, the family's tastes and preferences that affect their *practices*, such as having or not having a lawn, and the combination of rain and temperature that occurs.

The empirical test of this general hypothesis was through estimation by multiple-regression analysis of a demand equation

$$\log_{10} Q_{it} = a + bP_{it} + c\,EVT\text{-}R_t + dFCV_i + U_{it} \qquad (1)$$

where

Q_{it} is the quantity of water delivered to a single-family residence i in month t. Q is measured in 100 ft^3 per month.

P_{it} is the marginal price of water (i.e., the marginal block rate) implied by the quantity Q_{it}. P is measured in cents per 100 ft^3, adjusted to real prices rather than nominal prices by the Consumer Price Index.

$EVT\text{-}R_t$ is evapotranspiration minus rain for the month t. Evapotranspiration was computed by the Thornthwaite method as a function of recorded temperatures.[2] Rain was as officially recorded for Tucson. $EVT\text{-}R$ is measured in inches.

FCV_i is the full cash value of the single-family residence i as recorded by the county assessor in 1979.

U_{it} is the random error.

a, b, c, and d are the regression coefficients to be estimated.

Because the price variable cannot be directly observed, but is obtained by associating the observed quantity with the implied price from the increasing block rate schedule, it is possible that the real price associated with the individual's real demand curve is not initially obtained. This difficulty occurs because of the increasing block rate price schedule and the random error assumed to exist in the relationship. (It would also occur with a decreasing block rate schedule.) Our solution to this technical problem of statistical estimation and a detailed discussion of estimating demand where block rates exist are presented in the appendix.

The demand model is based on the reactions of individual householders to price and other surrounding conditions. The marginal price P_{it} was selected for use, rather than an average price, since it is the marginal price on the block rate schedule that intersects the demand curve. (See figure 5-1.) The individual's demand curve DD intersects the price schedule at marginal price P, resulting in water use of quantity Q.

The demand curve will shift to the right in the summer when evapotranspiration minus rain is higher than average. Demand will shift to the left when $EVT\text{-}R$ is less than average.

The variable FCV (full cash value) is included in the model for two reasons. First, one would expect FCV to be positively correlated with income. The higher a person's income, the less he or she would be expected to be responsive to price. Data on individuals' incomes could not be obtained. Second, a larger FCV would be

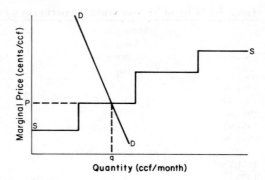

Figure 5-1. An individual's demand curve intersecting an increasing block rate price schedule in the second block. (ccf $= 100$ ft³)

expected to be positively associated with a large house and possibly a large yard. The larger the house and yard, the more water that would be used at any given price. Specific data on practices were not available for these individual residences.

The Sample

The Tucson Water Department's computerized record tape for 1976 was read using a computer, and the data for every twenty-fifth single-family residential customer was recorded. About 4,000 residences were selected. The addresses of these residences were matched to the data tape for calendar years 1974, 1975, 1977, 1978, and 1979, and the monthly consumption water data were added to those for 1976. The latest year for which data were available was 1979. The early year of 1974 was chosen as being two and one-half years before the political "furor."

In order to find the full cash value for each residence, the addresses selected from the Water Department tapes had to be matched with addresses from County Assessor tapes. Since the format of the addresses in the consumption file is different from that of the addresses in the assessor's parcel code file, a program was written to convert the addresses from the consumption file to a format that could be matched.

Complete data on consumption for the six years were matched with full cash value for 2,159 single-family residences. The 2,159 sets of observations were the basis of the regression analysis. The mean water use per single-family residence, based on this sample, is presented in table 5-2 by month, season, and year. These data are in 100 ft³ per month. The Water Department publishes summary data for all single-family residences in the Tucson system in gallons per day. When the summary data for all Tucson for 1978 and 1979 are converted to 100 ft³ per month, the comparisons between the summary data and the sample for mean 1978 in 100 ft³ are 13.44 versus 13.41, respectively. For mean 1979 the comparisons in 100 ft³ are 13.13 for the summary versus 13.22 for the sample. We conclude that the sample well represents the population of Tucson single-family residences. (Seven hundred forty-eight gallons = 100 ft³.)

Results of the Regression

An unsuccessful attempt was made to include all six years' data in the analysis—the thirty months before the events of July 1976 and the forty-two months thereafter. Two procedures were tried. The first was to add a 0, 1 dummy variable to the model described above. Zero was for observations before the "furor" in July 1976; 1.0 was for observations thereafter. While the estimated coefficients on all independent variables, including the dummy furor variable were highly significant statistically, the resulting estimate for the price elasticity of demand was unreasonably low. The combined effects of furor and price apparently were not being effectively sorted out. Still, the statistically significant coefficient on furor indicated a definite leftward shift in demand following the 1976 political upheaval.

Table 5-2. Mean Water Use per Single-family Residence, Tucson, 1974–1979
(100 ft³)

Month	Calendar Year					
	1974	1975	1976	1977	1978	1979
Jan.	11.45	10.65	11.08	9.85	9.66	8.62
Feb.	11.16	10.52	10.32	8.65	7.65	8.16
Mar.	13.70	11.96	11.08	9.98	8.89	8.81
Apr.	18.70	13.04	16.40	12.13	11.28	10.92
May	23.71	20.22	17.89	14.77	13.63	14.68
June	32.99	27.28	22.47	18.29	21.03	17.44
July	30.34	25.94	25.55	19.71	21.35	20.04
Aug.	22.21	23.05	16.93	15.65	15.00	14.94
Sept.	23.20	22.83	17.45	18.40	18.04	16.21
Oct.	15.66	18.43	12.24	12.94	14.27	15.29
Nov.	12.42	15.20	11.23	10.04	11.07	13.14
Dec.	11.13	11.84	9.95	9.80	9.05	10.41
Mean	18.89	17.59	15.22	13.35	13.41	13.22
1st 6-month mean	18.62	15.61	14.87	12.28	12.02	11.44
2nd 6-month mean	19.16	19.57	15.56	14.42	14.80	15.01
Nov.–Apr. Winter	13.09	12.22	11.67	10.08	9.60	10.01
May–Oct. Summer	24.69	22.96	18.75	16.63	17.22	16.43
	Fiscal Year					
	1974/75	1975/76	1976/77	1977/78	1978/79	
Mean	17.39	17.22	13.92	13.22	13.12	

Note: Random sample of 2,159 residences from Tucson Water Department records.

The second procedure was to divide the sample into two parts—the thirty months before July 1976 and the forty-two months thereafter. This procedure, if successful, would allow the slope of the two estimated demand curves to differ, as well as allowing a demand shift. Inspection of figure 4-2 in chapter 4 shows why neither procedure was successful. The sample data previous to July 1976 have almost no variation in price. The price schedule during that period (beginning with the new rate schedule of February 1, 1974) had only one step, did not vary between winter and summer, and was left unchanged until July 1976. Thus, these data cannot provide any information about the slope of the demand relationship in that period.

As a consequence of this lack of price variation, the successful regression was an estimate of demand based only on the forty-two months between July 1976 and December 1979. The estimated relationship is as follows:

$$\log_{10} Q = 0.820737 - 0.003750P$$
$$(0.010300) \quad (0.000406)$$
$$+ 0.040471 \ EVT\text{-}R$$
$$(0.000613) \tag{2}$$
$$+ 0.000005 \ FCV$$
$$(0.000000)$$

The standard errors of the estimates are shown in parentheses below each estimated coefficient. Each coefficient is highly significant statistically. The total explanation of variance is quite low ($R^2 = .102$), but one would expect it to be in a regression of this type based on individual householder behavior. The regression essentially estimates the behavior for the mean household. Individual deviations in demand could be quite large. Estimates of demand such as made by Billings and Agthe, where mean data are used for the regression, show a higher explanation of total variance, but this higher estimate is not meaningful.[3] In their procedure the individual variance has already been eliminated by taking the means before using the data in the model. In either case, it is the accuracy of the estimate of the price coefficient that is of importance.

The equation itself cannot efficiently be used for predicting quantities, since the dependent variable is in logs. The log form provides a better fit than does a simple linear model, since the log form compresses observations of large water use back toward the mean. However, the important estimate is the price elasticity of demand that is derived from the demand equation.

The price elasticity of demand (E_d) is defined as the percentage change in quantity purchased resulting from a percentage change in price.

$$E_d = \%\triangle Q / \%\triangle P \tag{3}$$

E_d is estimated as -0.256 at the mean price for the years 1976/77 through 1979.

An elasticity is termed *elastic* if its value is less than -1; that is, if the percentage change in quantity is greater than the percentage change in price. An elasticity is *inelastic* if its value lies between 0 and -1. An inelastic demand implies that a relatively large change in price will cause only a relatively small change in quantity demanded. The estimated price elasticity of demand for water in Tucson of -0.256 is considered inelastic, implying that a percentage change in price will change the quantity of water purchased by a lesser percentage. However, when we later examine this statistical evidence in light of the historical record of water quantities used and water prices charged, it becomes apparent that even such an inelastic relationship between price and quantity can have important effects on water consumption. But first, let us examine the strength of the relationship between quantity and one other component of the demand equation, evapotranspiration minus rain.

Evapotranspiration Minus Rain

Mean monthly deliveries of water per single-family dwelling are plotted against monthly *EVT-R* for the years 1974 through 1979. (See figure 5-2.) The dimensions are 100 ft^3 per month on the left vertical axis and *EVT-R* in inches on the right vertical axis. The relationship between the two sets of dimensions as plotted on the figure are arbitrary, but were selected so that the graphs of the two quantities are of similar heights.

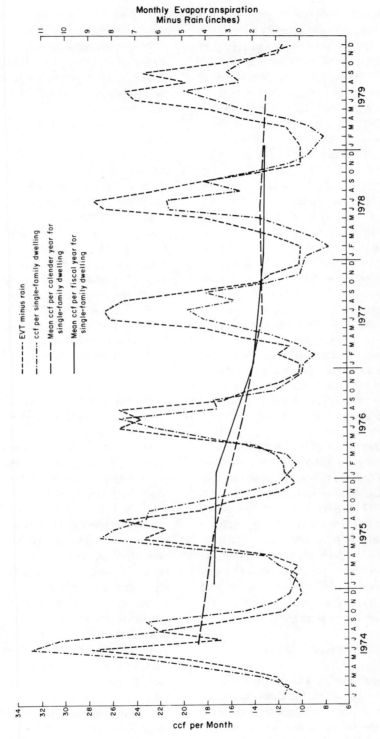

Figure 5-2. Monthly water deliveries per single-family dwelling, in relation to evapotranspiration minus rain, Tucson, 1974–1979 (sample data). (ccf = 100 ft³)

The correspondence between water use and *EVT-R* is striking. Water use closely follows the rise and fall of *EVT-R* in all years. In every year water use reaches a peak in either June or July, begins falling with the late summer rains, rises again in September as rains lessen but temperature remains high, and falls to a minimum in February. *EVT-R* is restricted to zero or above in figure 5-2 and in the regression model since we hypothesized that negative *EVT-R* would not further reduce water use. Apparently, however, water use is restricted further when a negative *EVT-R* occurs. See the winter months for 1977 and 1978 in figure 5-2 where *EVT-R* was plotted as zero rather than negative.

In fact, almost all of the *individual* variance in behavior is explained by *EVT-R* in the statistical demand equation. Yet, figure 5-2 clearly illustrates that while *EVT-R* follows approximately the same pattern in each year and has approximately the same mean value, the pattern of water use is substantially lower in recent years. Something has affected *average* behavior, whether it be price or conservation behavior causing a leftward shift in demand. As we discuss in detail in chapter 7, such a shift in demand could only be achieved by changing the practices of water use.

Mean monthly water usage for both the calendar and fiscal years are also plotted in figure 5-2. Based on the calendar year, mean monthly usage began declining as early as 1974 and stabilized after 1977. In terms of fiscal years, water use was almost stable through 1975/76, declined in 1976/77 and 1977/78, and then stabilized. The fiscal year view probably is most revealing, since water rates have generally been changed in the spring, or as in 1976 in the summer, and the following fiscal year would reflect any reaction to the new price schedule.

Water Prices and Demand

The regression model was estimated using monthly prices adjusted by the Consumer Price Index to 1966 = 100. In figure 5-3 the price schedules for the sample period are shown adjusted to 1979 dollars, the last year of the sample. It is evident from this figure that the adjusted price schedules were almost constant between 1977 and 1979.

The estimated demand curve $D_E D_E$ is also displayed in figure 5-3, adjusted to 1979 prices and the mean monthly usage for the years July 1976 through December 1979. Demand for the mean household intersects the supply schedules in those segments surrounding 1,500 ft^3 per month.

To illustrate the importance of using adjusted real prices rather than nominal prices, the marginal water prices per 100 ft^3 in the mean-use block are displayed in table 5-3 for each year since 1952 when a new rate schedule was introduced. Both nominal prices and real prices in 1979 dollars are listed. While the nominal price of water per 100 ft^3 in the mean-use block has increased from 10¢ to 88¢ (summer prices where relevant) from 1952 to 1981, real prices in 1979 dollars were 27.3¢ in 1952, rising to 79.0¢ in the summers of 1977 and 1978, and have declined substantially since, despite annual nominal price increases.

Implied Price Elasticities from the Historical Record

The price elasticity of demand (E_d) was estimated statistically only for the forty-two-month period following the political events of July 1976. Since that time, although the nominal price schedule has been raised each year and there has been continuing conservation publicity, there has been no additional political furor. We hypothesize that the demand curve has remained relatively static since the summer of 1976 without additional shifts because of basic changes in conservation behavior. E_d for this period was estimated statistically as -0.256. Shifts in demand apparently occur in reaction to noticeable political events in combination with significant change in the price schedule. Examination of table 5-4 provides a rough test of this hypothesis.

Column 2 lists the mean pumpage per capita in gallons per day for each fiscal year since 1965/66, as estimated by the Tucson Water Department. Columns 3 and 4 show the absolute

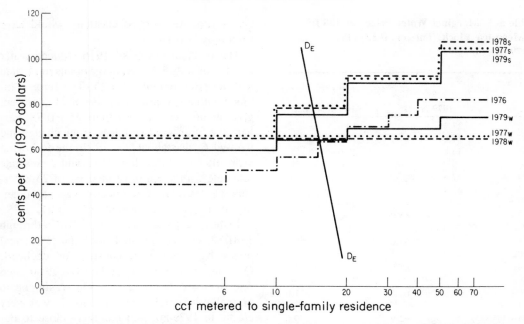

Figure 5-3. Single-family price schedules and estimated monthly demand, Tucson, 1976–1979 (1979 prices). (ccf = 100 ft³)

change and the percentage change, respectively, between each fiscal year. The marginal price in the mean-use block is listed for each year in column 5. All prices are adjusted to 1979 dollars. The absolute and percentage change in prices are listed in columns 6 and 7.

The implied E_d is computed as $\%\triangle Q/\%\triangle P$. These data are from columns 4 and 7. The implied E_d for each fiscal year is listed in column 8, and the directions of price and quantity change are noted in column 9. When price and quantity change in opposite directions, giving a negative implied elasticity, demand for water appears to be functioning normally. Of course, implied elasticities are based only on price changes, and in any given year weather or some other factor may have confounded the results.

A new price schedule was introduced in 1964/65 and remained in effect until 1970/71. The implied elasticity for 1966/67 is positive and thus apparently not rational; both real price and quantity used decline. However, the *apparent* positive elasticity could simply reflect a continued shift leftward in demand triggered by the nominal price increase in 1964/65 from 14¢ to 20¢ per 100 ft³, and an even larger real price

increase (see table 5-3). Once the effects of the price schedule change are stabilized, however, demand appears relatively elastic, increasing per capita quantities used significantly in response to real price declines. E_d appears to be in the range of −0.75 to −1.0.

In 1970/71 another price schedule increase was introduced. The nominal price increase was 5¢ and 25 percent of the former price (table 5-3). The real price increase was only 17.9 percent because of the real price decline since 1964/65 (table 5-4). The public apparently did not react immediately to the price schedule change. The implied elasticity was only −0.21, similar to that estimated by our demand equation. In the next two years, however, quantities fell even though the real price (but not the nominal price) fell slightly as well. In 1973/74 there is an anomaly. The real price declined somewhat, but the quantity of water used increased out of proportion with any other quantity change. The period 1970/71 to 1973/74 can only be classified as a period of neither economics nor politics.

Except for the year 1970/71 when a new price schedule was introduced, all real price changes since 1965/66 had been downward, but by relatively small percentages. In 1974/75 a new price

Table 5-3. Marginal Water Price per 100 ft^3 in Mean-use Block, Tucson, 1952–1981 (1979 prices)

(cents)

First year of schedule	1979 prices[a]		Nominal prices	
1952	27.3		10.0	
1959	34.8		14.0	
1964	46.8		20.0	
1970	46.7		25.0	
1974[b]	57.4		39.0	
1976[c]	63.8		50.0	
1977s[d]	79.0		66.0	
1977w		65.9		55.0
1978s	79.0		71.0	
1978w		65.6		59.0
1979s	76.0		76.0	
1979w		65.0		65.0
1980s	70.0		80.0	
1980w		62.3		71.0
1981s	70.2		88.0	
1981w		62.9		79.0
1982s	71.2		95.0	
1982w		60.0		80.0

Note: s = summer; w = winter.

[a]Consumer price indices (CPI) for 1952 through 1980 taken from "Working Data for Demand Analysis," Economic Research Service, USDA, revised June 1981. CPI for 1981 and 1982 are from *The Manufacturers Hanover Economic Report* (350 Park Avenue, New York, July 1982). The final two years have not yet been revised so as to make a fully consistent series.

[b]Between the years 1973 and 1974, the real jump in 1979 prices was from 40.8¢ to 57.4¢ (16.6¢).

[c]Between 1975 and 1976, the real jump in 1979 prices was 52.6¢ to 63.8¢ (11.2¢).

[d]After the real price jump of 15.2¢ between 1976 and 1977 summer prices, real prices have declined.

schedule was introduced with a 40.6 percent increase in real prices. Little political activity accompanied the price rise, however, and while water use dropped precipitously, a price elasticity of only −0.32 was implied. In the following year, 1975/76, a small real price fall brought on a quantity rise. The implied elasticity is of the correct sign and apparently relatively elastic. However, there is a problem with the data series of estimated per capita pumpage in this year; the absolute value of the implied elasticity probably is much larger than in reality. Up until this year the Tucson Water Department had assumed 3.7 people per service hookup. In 1975/76, they changed their assumption to 3.5 people per service, causing a discontinuity in the series. If the original assumption remained

unchanged, the implied elasticity would have been only −0.10.

The political events of 1976 reduced water use, but by only 3.5 percentage points more than had the price rise of 1974/75. The furor combined with a real price increase of 21.3 percent gives an implied price elasticity of −0.78. Perhaps the furor was two-thirds of the change and the real E_d was about −0.25. In the following year, after the furor died down and a new rate schedule with a real price increase of 23.8 percent was introduced, the decline in water quantity implied a price elasticity of −0.21.

In the recent four fiscal years 1978/79 through 1981/82, despite nominal price increases, real prices have remained constant or declined. Quantities fell slightly after the 1978/79 nominal price increase, but have begun increasing again as real prices fall. The implied E_d was very elastic in 1979/80, and relatively close to our statistical estimate in 1980/81. The most recent fiscal year, 1981/82, saw almost no real price change, but a sizable increase in per capita use. News about water has been continuous, but no crises have occurred.

Of the most significance in this historical summary is that the implied elasticity is of the correct sign in eleven out of sixteen years, indeterminant in two years, and incorrect only in three years. We conclude that decreasing real prices encourage increasing water use. Increasing real prices encourage less water use, but the static demand curve, at least since 1974/75, has been relatively inelastic. Major decreases in water consumption per capita occur only where a major price increase is accompanied by major public awareness of the action surrounding the passage of the increased price schedule.

Water Price—A Summary

Contrary to the original assumptions of the Tucson water managers who saw price increases simply as a means of raising additional revenues, price is an important component affecting consumer behavior. Increasing *real* prices reduces per capita water use, and decreasing *real* prices brings about higher use. The public is not

Table 5-4. Water Pumpage, Water Price, and Implied Price Elasticities, Tucson, Fiscal Years 1964/65 to 1981/82

Fiscal Year (1)	Q^a (2)	$\triangle Q^b$ (3)	$\%\triangle Q$ (4)	P^c (5)	$\triangle P^d$ (6)	$\%\triangle P$ (7)	$E_d{}^e$ (8)	Directions of change (9)
1964/65								
1965/66	179.2			46.0				
1966/67	176.3	− 2.9	− 1.6	44.7	− 1.3	− 2.8	+ .57	$P\downarrow Q\downarrow$
1967/68	181.6	+ 5.3	+ 3.0	43.5	− 1.2	− 2.7	− 1.11	$P\downarrow Q\uparrow$
1968/69	186.7	+ 5.1	+ 2.8	41.8	− 1.7	− 3.9	− .72	$P\downarrow Q\uparrow$
1969/70	194.2	+ 7.5	+ 4.0	39.6	− 2.2	− 5.3	− .75	$P\downarrow Q\uparrow$
1970/71	187.0	− 7.2	− 3.7	46.7	+ 7.1	+ 17.9	− 1.2	$P\uparrow Q\downarrow$
1971/72	184.2	− 2.8	− 1.5	44.8	− 1.9	− 4.1	+ .37	$P\downarrow Q\downarrow$
1972/73	170.8	− 13.4	− 7.3	43.4	− 1.4	− 3.1	+ 2.35	$P\downarrow Q\downarrow$
1973/74	204.6	+ 33.8	+ 19.8	40.8	− 2.6	− 6.0	− 3.30	$P\downarrow Q\uparrow$
1974/75	177.7	− 26.9	− 13.1	57.4	+ 16.6	+ 40.6	− .32	$P\uparrow Q\downarrow$
1975/76	189.4	+ 12.0	+ 6.8	52.6	− 4.8	− 8.4	− .80	$P\downarrow Q\uparrow$
1976/77	157.9	− 31.5	− 16.6	63.8	+ 11.2	+ 21.3	− .78	$P\uparrow Q\downarrow$
1977/78	150.1	− 7.8	− 4.9	79.0	+ 15.2	+ 23.8	− .21	$P\uparrow Q\downarrow$
1978/79	147.5	− 2.6	− 1.7	79.0	0.0	0.0	f	$P\rightarrow Q\downarrow$
1979/80	154.4	+ 9.5	+ 6.4	76.0	− 3.0	− 3.8	− 1.68	$P\downarrow Q\uparrow$
1980/81	158.9	+ 4.5	+ 2.9	70.0	− 6.0	− 7.9	− .37	$P\downarrow Q\uparrow$
1981/82	165.6	+ 6.7	+ 4.2	70.2	+ 0.2	0.0	f	$P\rightarrow Q\uparrow$

Note: Horizontal lines indicate years in which a new price schedule was introduced. Quantity for 1964/65 was not available.

[a]Pumpage per capita in gallons per day.

[b]Change in pumpage between years.

[c]Marginal price per 100 ft^3 in the mean-use range. Summer prices are used from 1977/78 to 1980/81. Previously, there was no difference between summer and winter prices.

[d]Change in prices between years.

[e]The implied price elasticity of demand: $\%\triangle Q/\%\triangle P$. Oil prices are in $1979.

[f]Undefined.

fooled by *nominal* price increases, despite its general distaste for any increase at all.

But price is not the only variable affecting behavior. The demand for water tends to shift leftward when an increasing price signal is accompanied by preachment and politics that reinforce that price signal. Since weather is such an important variable affecting water use, even the signals of price, preachments, and politics will only affect use as basic water practices can be changed (see chapter 7).

Since price is an important component affecting water use, the question of how to price water becomes a crucial issue. If a water utility faces increasing marginal water costs, which is the same as saying that water is becoming increasingly more scarce, all water delivered should be priced at its marginal cost. If this rule is followed, consumers will receive a correct signal about the relative scarcity of water and react accordingly. To preach water scarcity while hiding the real cost of water sends conflicting signals.

Tucson is a perfect example of conflicting signals confusing the issues surrounding water use. While water scarcity and the need for water conservation are preached, and a relatively innovative pricing policy has been introduced, lack of understanding of marginal pricing policy along with a lack of political will to face unpleasant issues keeps water prices artificially low. Prices are still set using historical average costs, and no value at all is placed on the current single source of water in the underground aquifer. The following chapter takes up the political problems of water policy.

Notes

1. Per capita consumption is an estimate based on an assumed number of people per service connection. Until early 1982, the lowest per capita use in Tucson was estimated to be 139.5 gallons per day, a usage that was widely cited as showing what conservation advocacy could do. Per capita use estimates were recently revised upward for fiscal year 1975/76 onward, based on a new, lower estimate of the service population. The break between fiscal years 1974/75 and 1975/76 is arbitrary, and creates some problems in comparing water usage before and after those dates.

2. Ven te Chow, *Handbook of Applied Hydrology* (New York, McGraw-Hill, 1964) p. 11.25.

3. R. Bruce Billings and Donald E. Agthe, "Water Price Elasticity in the Case of Increasing Block Rates: Some Evidence from Tucson, Arizona" *Land Economics* vol. 56 (February 1980) pp. 73–84.

6

The Political Feasibility of Raising Consumer's Water Costs

The Political Puzzle

The reason and wisdom of economics are all too often politically insupportable. The marginal-cost pricing of water advanced in chapter 4 implies stiff rate increases. Increasing individuals' costs—even if such increases are economically efficient, and equitable in the sense of eliminating subsidies—is not attractive politically, as contemporary attempts to cut human services, share entitlement programs, and raise taxes illustrate. Handing out disadvantages ordinarily activates opposition, and in the long run any policy, or politician for that matter, must generate a favorable balance of approval. No matter how efficient an allocation of resources a policy may offer, politicians tend not to take economists' prescriptions if the side effect of the medicine is political suicide.

Policies that create sharp increases in cost of public services are difficult to initiate, legitimize, and implement.[1] If a burden can be heaped only on a minority of people, particularly persons in a group that can be isolated or portrayed as undeserving, then increasing costs might be politically acceptable. However, rate increases usually affect all users. If there were counterbalancing positive advantages given directly to particular interests, strong policy champions might emerge. However, the rewards from rate increases are likely to be amorphous and remote, such as the financial health of the municipal service system. Such benefits are similar to public goods: they are not divisible into particular shares that can be claimed by individual or group recipients. Users do not see themselves as "the beneficiary." Political support is most easily motivated by benefits such as subsidies or tax breaks that make those who receive them relatively better off. It is more difficult to persuade people to suffer particular costs and forgo present benefits in favor of some future gain that accrues generally to everyone or to some future generation than to gain support from those who would receive a subsidy today.

Water, sewer, street maintenance, and similar municipal governmental functions are ordinarily low-visibility and low-interest issues that seldom become politicized. When these matters do become public controversies, it is nearly always detrimental to persons in authority. Citizens expect government to perform these functions in

69

the same way they might expect a dental clinic to perform. While they suppose that the techniques are complicated, they really do not want to hear about them. If more than the expected pain is involved, they conclude that the job is not being done right and simply find a new dentist. The patient does not expect the dentist to ask about tastes and preferences, but to do what is considered professionally best. Certainly the client does not expect to be told that past treatment was mistaken and that considerable discomfort must be borne while matters are straightened out.

For municipal politicians, the best service issue is no issue. Municipal politicians have every reason to leave service issues to the relevant government agencies, expecting well-run and scandalfree operations in exchange for agency autonomy from political meddling. Ordinarily, politicians can only get into trouble when events raise services to the public consciousness. For example, Jane Byrne's challenge of the Chicago machine profited considerably when the great Chicago snowstorm put commuter trains out of commission.

The incentives for municipal service agencies are for conservative, cautious management. Since politicians who get involved in the business of municipal services are likely to be angry politicians who clear people out when they decide to clean things up, political involvement is avoided. Decisions on such matters as proposed rate increases most likely are incremental. Innovative ideas are experimental, and while they may promise large changes for the better, the risks of unanticipated negative consequences are larger than they are for small changes. Service managers tend to follow tried-and-true rules— doing things the way everyone in the business does them. Should problems arise, being professionally sound provides managers with a powerful defense.

Ordinarily it is not politically feasible to make large changes and impose sharp new costs in municipal services rate structures. Yet in 1976 the City of Tucson Water Department recommended and the City Council adopted a radically different rate structure that hurt customers directly. More than six years later the essentials of the changed rate structure are still in place. What kind of political dynamics underlay political events that so sharply diverged from conventional political wisdom? Are there lessons in the Tucson experience for political actors in other cities facing similar water problems? What are the implications of Tucson events for the political acceptability of marginal-cost pricing?

The policy-making process can be divided into three broad stages: (1) policy initiation and formulation; (2) policy legitimation; and (3) policy implementation. The following sections discuss the difficulties encountered by policymakers in each of these stages, show how these difficulties were overcome in Tucson, and draw some general lessons from the Tucson experience.

Policy Initiation and Formulation

Initiatives that are actively and seriously considered by governmental institutions make up the broad institutional agenda.[2] Within this broad agenda of priority items there are subsets of issues, each bundle of which is handled differently by government. One such agenda is bureaucratic, and its treatment is determined by established practice and routine. Handling of the bureaucratic agenda is characterized by absence of open or expressed conflict; most active participants are administrative agency officials. Normal municipal service decisions are habitual decisions and are made without requiring anything like a new initiative or reformulation of the issue. Items such as general budget, marginal expansion plans, and normal rate increases stay on the bureaucratic agenda for management decisions and are simply ratified or legitimated by city councils.

The political agenda differs from the bureaucratic agenda in that the issues are conflictual and require treatment by elected politicians whose job it is to deal with clashes of values. Nonincremental decisions, especially those that threaten the wide distribution of unaccustomed costs, cannot be contained in a bureaucratic arena. Such policies require redefining the issue and getting it on the political agenda.[3] In their book on agenda-setting, Cobb and Elder describe circumstances that lead to issue expansion, that is, moving an item from the bureaucratic to the political agenda and involving other than agency participants in decision making.[4] When some participants are dissatisfied with the disposition of an issue that is normally handled routinely, those participants may invite conflict in hope that the policy change they desire will fare better in a broader political arena.

In view of public opinion about water availability and use in Tucson in 1976, it is difficult to imagine why participants in favor of an innovative policy would want to politicize the issue and expand it to a more public arena. A public opinion survey conducted that year indicated that Tucsonans had not really considered the water allocation problem in the Tucson Basin and were not prepared to require any class of user to cut back. Residents were asked, "(If) the Southwest may eventually have to set priorities among various water users, including farms, cities, energy, etc., in your opinion, should each of the following users get more, the same, or less water in the future?" Among those classes of water users shown in table 6-1, not one user group was chosen by a majority of Tucson respondents to receive less water in the future. In fact, more than 80 percent of Tucsonans sampled felt that all users, with the possible exception of recreational users, should receive the same amount or more water in the future.

Since the participants in broadening the political arena could not reasonably have anticipated public understanding and support, they must have had other motivations. It is helpful to look more closely at the risks and rewards of the normal participants, that is, the Water Department officials and the members of the City Council, who routinely approve utility decisions.

Water utility officials normally do not want to raise public interest in water allocations and cost. The best way for a water utility to remain noncontroversial and to fulfill public expectations is to provide whatever water is necessary to meet demand. As the demand for water has grown, water systems have been enlarged to meet the increased demand.[5] Partly because water is regarded as a necessity, and partly because the enlargement of a water system has very little effect on the cost of supplying a unit of water, the need for enlarging a water system, in most cases, has been left unquestioned.[6]

This emphasis on supply solutions to water management needs has been reflected in the rate structures generally employed by water utilities. As explained in chapter 4, the price of water has been determined by dividing the total system costs by the amount of water delivered, usually categorized by class of customer. This average-cost pricing has tended to promote system expansion by encouraging overinvestment in system capacity.[7] Decreasing block rates, where water prices decrease as consumption increases, are currently used by an estimated 94 percent of all municipalities.[8] In pursuing these management strategies, utilities have been able to give the public what it wants—water that is cheap and plentiful and water service that does not create conflict among users. There has been no incentive to enlarge the issue by creating controversy with innovative proposals for changing rates.

By 1975, however, the traditional approach to water utility management, although firmly entrenched among water professionals and supported by public opinion, had become more dif-

Table 6-1. Priorities of Tucson Residents for Future Water Allocations Among Water Users

Use or user	Percentage of residents favoring:			Number in sample
	More water	Same amount	Less water	
Electrical energy production	37.5	53.9	8.6	128
Irrigated agriculture	39.7	45.6	14.7	136
Industry and manufacturing	14.1	70.3	15.6	128
Municipal and residential	23.2	65.2	11.6	138
Water-based recreation	17.3	46.0	36.7	139
American Indians	29.7	63.3	7.0	128

Source: Four Corners Project survey data. See Helen M. Ingram, Nancy K. Laney, and John R. McCain, *A Policy Approach to Political Representation: Lessons from the Four Corners States* (Baltimore, Johns Hopkins University Press for Resources for the Future, 1980).

ficult to pursue. Greater consumer demand for water, difficulties in expanding supplies, more stringent water quality standards, and rapidly escalating energy costs for pumping and transporting water all served to complicate the process of providing inexpensive water to meet all demands.

Faced with these pressures, waterworks professionals began to entertain some new ideas about the best way to serve their customers. By 1975, articles on innovative rate schedules had begun to appear in the trade journals.[9] These studies represented a growing interest in the purposes as well as in the structure of water rate schedules. The motivation for these studies has been the idea that water rates should more closely reflect the cost of serving a customer. In the warmer and drier areas of the United States, the large amount of water used for watering lawns and gardens results in a high peak demand for water during the summer months. Since a large proportion of many water utilities' costs results from the need to meet this peak demand, many waterworks managers had become interested in designing rate structures that would recover the cost of providing for the peak from the customers responsible. By 1976 a number of water utilities such as the Fairfax County, Virginia, Water Authority, the Washington (D.C.) Suburban Sanitary Commission, the Denver Water Board, and the East Bay Municipal Utility District (California) had begun to assess the effects and feasibility of novel rate structures.

This trend represented a significant departure from the traditional approach to rate structures long followed by American waterworks professionals. The "cost of service" criterion for customer charges, a well-established practice in most American business, was very much a new idea to the service-oriented water utility industry. The new rates offered water utility managers a strategy for dealing with high peak demands and large costs imposed on the system by what they viewed as only some customers. Despite the equity and conservation implications of these innovative rate structures, the primary motivations of waterworks professionals in examining new rate structures were practical management

problems. The cost-of-service trend was limited at this time to a willingness to consider and study new rate structures. Implementation of these ideas had not yet occurred in a real waterworks setting, and the verdict was far from in about the advantages of rates based on the cost of service.

The situation in the Tucson water utility did enhance the possibilities of issue expansion in favor of innovative rate making. First, there was the real financial difficulty already described. Second was the importance of the presence of a number of unusually outspoken, bright, and aggressive staff members knowledgeable about the literature and committed to managing water better. Finally, the periodic justification and system evaluation performed by the city's rate consultant provided a timely opportunity for new proposals.

The head of the Water Department, Frank Brooks, had already proved his political courage by questioning the benefits of the popular Central Arizona Project (CAP) to the city in the absence of agricultural water use reform.[10] A key staff member, Jerry Wright, held a Ph.D. in hydrology from the University of Arizona. Wright saw water in systematic terms, and was deeply concerned that the limits of water resources in the Tucson Basin be recognized and included in planning.[11] Two other staff persons, Stephen Davis and John McGill, brought technical skills to the department and were interested in increasing the water data base and in using computer techniques. When more advanced analytical techniques were integrated into the water planning process, the capital needs of the system were revealed in stark terms. The staff came to believe that the financial basis of the department was seriously threatened and that extraordinary action needed to be taken, such as adopting the cost-of-service approach discussed in the current literature. The periodic review by John Carollo Engineers offered an opportunity to explore such an approach and develop a detailed rationale for it. With the City Council's permission, Carollo Engineers was asked to develop these and other concepts, such as increasing block rates, in its 1976 report. The Tucson Water Department staff would not have ventured to recommend such large changes, however, had it not been for the support of the New Democrats on the City Council.

The real puzzle of the Tucson water rates case is why the majority of the Tucson City Council participated in putting innovative water rates onto the political agenda. Conventional political wisdom would suggest that the matter of innovative water rates was a poor political issue; the rational political response would have been to insist that the water utility solve its problems as best it could in the habitual, incremental manner. Enlarging the issue from simple bureaucratic problem-solving risked displacing other, more profitable issues, and, more seriously, risked generating constituency disapproval. The five propositions discussed below seem best to explain the council's willingness to take on these risks.

Risks of Inaction

First, the risks of inaction were greater than the risks of action. The Tucson water utility was having trouble delivering water to some areas at peak hours on peak days. If peak use continued to climb, some areas would not have service. Customer response was bound to be critical. To secure more water, new capacity had to be developed, and the utility lacked funds.

Critics have questioned whether the crisis was in fact real.[12] The leader of the recall campaign, John Varga, believed that the council was manipulated by the Water Department staff and that money actually was being raised for bureaucratic aggrandizement, including the purchase of private water companies.[13] More to the point, however, the City Council majority did not need to take the *specific* actions they did (lift charges, increasing block rates, system development charges) in order to respond to crisis conditions.

Initial Tolerance of Risk

Second, political actors are most likely to take risks at the beginning of their political careers. When they first come to office, politicians are anxious to make names for themselves.[14] Their own political agendas are not established, and they are open to opportunities for distinguishing themselves. Witness that President Carter's highly risky hit list of water projects to be eliminated from the budget was the product of a transition period. Two of the New Democrats were elected to the Tucson City Council in 1973 and two in 1975, making a majority. The rate question presented a chance for this new governing group to act, although if they had had time to explore alternatives, they probably would have chosen some other area in which to innovate. Further, new politicians do not share the conventional issue context. The New Democrats were not schooled in thinking that water was a routine issue requiring only incremental change.

Greater Acceptance of Shared Risk

Third, the more a risk is shared, the more actors will be ready to accept it. The innovative rate schedule adopted in Tucson was a collective rather than an individual product. New Democratic council members had a high level of cohesion. Council members and their aides worked together with the staff of the Water Department. Individuals reinforced one another's view of the correctness of proposed action. Since they heard no words of caution from each other, danger was discounted. It seemed unlikely that all the right people could be wrong.

Uncertain Risks Are Discounted

Fourth, perception of risk depends upon information, and risks that are unspecified and uncertain are likely to be discounted in favor of risks about which there is more information. There was a distinct imbalance in the kind of information council members received—mostly technical and very little political. Council members were confronted with an extraordinarily well-documented case for a water crisis requiring immediate action from the Tucson Water Department. Council members moved quickly into the details of proposed rate changes (consideration of how steep increasing block rates should be, designation of lift zones, and so forth). The Republican mayor remained in the background, raising objections toward the end of the process, but avoiding association with the rate plan or any alternative to it. One can only speculate about whether the strategy of sitting back and letting the New Democrats get themselves into trouble was purposeful.

There were few early overt warnings of the political mine field that lay ahead. A large rate increase in 1974 had not produced any outcry from the public. There was no controversy over rates in the newspaper before the recall was initiated. Neither the Water Department staff nor the City Council members anticipated the large rate increases that some users in high lift zones would receive. The New Democrats on the council were warned of the political risk by their colleague and County Supervisor Ron Asta, but his word of warning came too late, seeming to contradict all the council's careful technical work.

In comparison with the hard physical facts, political risks seemed very speculative. The council members were still in the flush of election victory, which they believed was a mandate for controlled growth. The next election was distant, and by then citizens could be expected to have accepted rate changes. Further, a recall election had never been held for a cause other than malfeasance.

Tolerance of Risk Greatest for Core Values

Fifth, the more central an issue is to one's value system, the greater the risk one is willing to take. While the New Democrats were motivated by a number of forces, it is clear that their preservationist conservation values were important. Profligate use of water in a desert environment was repugnant to them. Making users pay the cost of service appealed to their ethical sensibilities. Raising water rates sent a conservation message to the public. While they thought the action would not be popular, they knew it was right.

The New Democrats on the City Council wanted to teach water users a lesson. If users were shocked, that was all the better. Councilman Robert Cauthorn, a professor of economics, believed that people were most likely to respond to pocketbook pressure. He was quoted as saying that he wanted to get the attention of "Mr. and Mrs. Crabgrass."[15]

While the council majority's rate-making actions can be described as a calculated risk, it is also possible simply to acknowledge that these politicians made a political mistake. In that they would not be alone. James Coleman has examined issue expansion in the context of local politics. He notes that often a series of errors committed by the leaders of the community activates the passive and disinterested public.[16] In this case the New Democrats provided an opportunity for outside mobilization by policy entrepreneurs Varga and Fitzgerald, the leaders of the 1976 recall campaign. Varga had been distrustful of the New Democrats since they had replaced the city manager. He had been looking for mistakes and blunders. After water users received their first bills with raised rates, an advertisement asking people to carry recall petitions gleaned 150 responses.[17] The subsequent defeat of the New Democrats was detailed in chapter 2.

Policy Legitimization

Legitimization in the policy process is the stage at which actions become definitive and authoritative. What had been a proposal or plan becomes policy in fact. Legitimization takes place through proper procedures in appropriate political arenas. Which arenas become involved depends upon the issue. With an expanded, conflictual policy, which involves more arenas tha does routine policy, decisions made in one setting are appealed to other settings. In such cases, the policy in question can be said to be legitimized when some authority hands down the final word, that is, when no further appeals occur.[18] The processes of gaining consent varies from one institutional setting to another, but building a majority is a common procedure by which policies are legitimized.

In the Tucson water case, where a decision on water rates made in the City Council was appealed to and rejected by the voters, the rate decision would appear to have lost all chance at legitimization. It would seem that any policy that resulted in the overwhelming defeat of its initiators at the polls would quickly be repudiated by their successors. Yet, what in fact occurred was a reconciliation. Staunch critics of increased rates eventually agreed that the policy was necessary. We explore how this reconciliation occurred in order to understand what happened in Tucson, as well as to learn more generally about the process by which political systems come to legitimize unpopular policies.

Water, it must be admitted at the outset, was not the only cause of the New Democrats' defeat. Political style was important. The New Democrats were viewed as high-handed, arrogant, and disinterested in public participation.[19] There was a more general adverse reaction to these political mavericks and their lack of enthusiasm for facilitating growth. For those in certain segments of the Tucson political and business community who viewed the New Democrats as a real threat, the recall movement was a fortuitous opportunity to oust them from office. Almost immediately after the new water rates were implemented, dire predictions of economic calamity abounded. Spokespersons for the home building and construction industry made loud claims that high water costs in general and the system development charge in particular would make it too expensive for people to move to Tucson. The Good Government League quickly expanded the debate from the subject of water into other controversial areas. The uncertainty and frustration of the public created by the political crisis surrounding the new water policy was capitalized on by the political opponents of the New Democrats. The following analysis of the election shows that water rates were directly related to electoral results, although they do not explain all of the variance in electoral response.

The New Democrats' faction lost on a city-wide basis by margins of approximately 21 to 30 percentage points. Turnout was heavy for a recall election. Election officials had predicted that between 20 and 30 percent of the city's registered voters would go to the polls, but 37.2 percent cast ballots. The exact extent of the contribution made by the water issue to the New Democrats' defeat requires analysis for which we have inadequate data. Ideally we should cor-

relate each consumer's rate increase and attitude toward the no-growth issue in general to the vote for recall. Since no such records exist, we look at voting in geographic areas where users were likely to have had increased water bills. Increasing block rates penalized all large users, and homes with green landscaping are scattered around town. However, there is a concentration of affluent residents with large yards in the Tucson foothills. Foothills residents also had to pay lift charges, a controversial part of the water rate increase. While not a perfect indicator of the rate burden, an analysis of voting by lift zone provides some insight into the electoral magnitude of the water issue.

We turn first to the relationship of rate increases to voter turnout. The heaviest turnouts were in wards 2 (43.6 percent) and 4 (40.5 percent) on the city's far east side. Those two were the only wards with more than one lift zone, and in some precincts in those wards, the new politics faction was battered by margins as large as six to one. Voter turnout may also be examined specifically by lift zone (table 6-2). It is clear that turnout was considerably higher in those areas of the city having one or more lift zones.

It could be argued that political party affiliation is closely related to residence, and that it was more important than were rate increases in affecting voter turnout. Turnout results could reflect the influence of party registration and the higher propensity for Republicans than for Democrats to turn out at the polls. In Tucson, relatively larger percentages of Republicans reside at the higher elevations than at lower elevations. The percentage of registered Democrats and Republicans is presented in table 6-3 by lift zone. Clearly, a relatively higher percentage of Re-

Table 6-2. Percentage of Voter Turnout by Lift Zone, 1976 Tucson Recall Election

	Lift zones[a]					
	0	1	2	3	Mean of 1, 2, & 3	Mean of all
Turnout	34.0	41.5	40.5	40.6	41.1	37.2

[a]The number of the lift zone corresponds to the number of lift charges.

Table 6-3. Percentage of Democrats and Republicans by Lift Zone, Tucson, 1976

Lift zones[a]	Democrats	Republicans	Excess of Democrats over Republicans (percentage points)
0	63.2	27.3	35.9
1	56.2	33.6	22.6
2	53.4	37.4	16.0
3	49.2	39.3	9.9

[a]The number of the lift zone corresponds to the number of lift charges.

Table 6-4. Average Percent Margin of Victory for all Winning Candidates, by Lift Zone in Each Ward, Tucson, 1976

Lift zone	Ward					
	1	2	3	4	5	6
0	13.1	11.3	10.5	—	10.5	11.6
1	38.9	31.5	—	23.4	28.5	24.8
2	—	35.8	—	35.8	—	—
3	—	47.1	—	44.9	—	—

Notes: The data for each ward are interpreted in the same way as the following example for ward 1: Of the total vote in ward 1, the average percentage margin of victory for all candidates was 13.1 percent in lift zone 0 and 38.9 percent in lift zone 1.

Winning candidates were Richard Amlee, James Hooten, Cheri Cross, and Schuyler Lininger.

Table 6-5. Average Percent Margin of Victory for Each Winning Candidate, by Lift Zone, Tucson, 1976

Lift zone	Richard Amlee, elected from ward 2	James Hooten, elected from ward 3	Cheri Cross, elected from ward 4	Schuyler Lininger, elected from ward 6
0	3.4	0.8	8.7	13.2
1	30.5	9.5	32.8	39.2
2	36.7	17.7	41.8	46.9
3	50.8	21.4	53.6	56.7

Note: The data for each ward are interpreted as in the following example for ward 2: The average percentage margin of victory for the candidate elected from ward 2 was 3.4 percent in lift zone 0, 30.5 percent in lift zone 1, 36.7 percent in lift zone 2, and 50.8 percent in lift zone 3.

publicans is found in lift zone 3 than in lift zones 2, 1, or 0. Percentage turnout was almost equal in lift zones 1, 2, and 3 (table 6-2). Thus, we conclude that turnout was more strongly a function of whether one had to pay any lift charges than of one's party preference.

The relationship of lift zone to actual vote is displayed in tables 6-4 and 6-5. Candidates are nominated and seek election from a particular ward, but voters in all wards vote on the candidate in each ward. In 1976 council members from wards 1 and 5 did not face recall. Therefore, table 6-4 contains results of all six wards, but only the results of the four contested ward seats are shown in table 6-5.

It is clear that the margins of victory for all winning candidates together (table 6-4) and for each winning candidate alone (table 6-5) are significantly different by lift zone. Indeed, in each table the percentage margin of victory increases as one moves from lift zone 1 to 2 to

3. The relatively lower margin of victory for the candidate from ward 3 (table 6-5) is explained by the fact that the incumbent who had represented that ward resigned. Thus, ward 3 had a wide-open race between James Hooten and other candidates unaligned with the New Democratic movement. If, however, the percentage margins of victory are combined for the candidate receiving the endorsement of the Citizens' Recall Committee and the candidate receiving the endorsement of the Good Government League, the organizations orchestrating the recall, the results are similar to those for the other contests, that is, 36.2 percent, 42.1 percent, and 47.9 percent in lift zones 1, 2, and 3, respectively.

There is other evidence that indicates that voters did not specifically reject water rates policy, but rather the way in which the policy was introduced and the persons who introduced it. The City Council, dominated by the New Democrats, commissioned a survey of a probability sample of water customers.[20] The results, not reported until after the election, indicated surprising acceptance of the idea of increasing block rates (70 percent of respondents said they were a good idea) and system development charges (73 percent in favor) among the 1,034 customers sampled. A majority of those questioned (57 percent) even agreed that all lift costs should be paid by those living in high lift zones. The survey did not tap consumers' perceptions of the severity, unpredictability, and inequity of the

New Democrats' rate increases. Also, there is considerable difference between survey respondents' approving these ideas in the abstract and liking their own water bills. The water bills apparently were more relevant to the election. Nonetheless, the survey indicated that once the offending council persons were swept away, voters perhaps could become reconciled to the new water rates policy.

However, the recall candidates ran on the platform of rolling back water rates. Why were only marginal changes made? The three following explanations seem controlling.

First, risk of action was greater than risk of inaction. The Tucson Water Department continued to provide the water management expertise and controlled much of the information that came to newly elected council members. New members were laypersons, recruited by the Citizens' Recall Committee to mirror the common citizen, who had no technical background and little political experience. Faced by reports of the Water Department, the Citizens' Water Advisory Committee, and the new consultant claiming that dire consequences would follow if rates were lowered, the council could not bring itself to roll back rates. It did, however, drop the most offensive provisions—the lift and system development charges. These charges were not critical to meeting the Water Department's financial obligations.

Second, the political costs of legitimizing unpopular policy was borne by others. The enduring Tucson politicians, particularly Mayor Lewis Murphy, came out of the Tucson experience with a policy in place that alleviated the financial emergency of the Water Department without having to pay a political price for such action. Costs were entirely borne by dispensable actors. The previous New Democratic council members bore the brunt of having handled things badly. Postrecall council members were for the most part ineffectual and were defeated when they came up for reelection in contests where water was not much of an issue. Frank Brooks stepped down from directorship of the Water Department, and a new director concerned with restoring the department's apolitical image took over.[21]

A mechanism was fashioned in the fall of 1976 whereby the costs of future water rate increases could be deflected from elected politicians. A Citizens' Water Advisory Committee (CWAC) was appointed by the city manager and made a permanent body in 1977. From the beginning the CWAC has aimed to be apolitical and to give sound, technically based policy advice to the city.[22] The council elected in 1976 relied heavily on the CWAC findings in deciding not to roll back rates. Since then, the CWAC has come to play a stalking-horse role, recommending higher rates and other changes more advanced than those suggested by the Water Department or City Council. The council at first resists, and complains of adverse impact on water users, but at length accedes to most of the CWAC's recommendations.

Third, additional policies provided opportunities for politicians to take political credit. The Beat the Peak campaign was (and is) a low-cost, low-risk way in which politicians could appear to be working on the water problem. It had the advantage of being noncoercive. In the words of Mayor Murphy,

We explained to the people that we would not dictate to them about water use habits, but we did point out things that made sense. Why have a 16 lane speedway just for rush hour traffic? With that common sense explanation, and the incentive of summer rate increase, everyone had every reason to cooperate.[23]

The campaign was a low-budget affair. It could scarcely fail, since increased water rates and greater public awareness were lowering people's water use. Whether the peak has been lowered more than could be reasonably expected from an overall decline in use is a matter of dispute, as was discussed in chapter 4. But the campaign ultimately placed the blame for increasing water rates not on the government, but on the public that refused to beat the peak. When peak use went down, officials of the city government took credit for a program that worked.

Policy Implementation

Implementation is relatively new as an area of public policy research, but researchers have discovered it to be the most problematic policy stage.[24] The determinants of policy success— or, more commonly than policymakers would like, the determinants of policy failure—are numerous and complex.[25] Among the most important of these factors is target-group compliance. The persons toward whom the policy is aimed must not pursue strategies of continued resistance. Instead they must begin to modify their behavior in the direction the policy intended.

There are numerous examples of policies legitimized in the political process that are not accepted as final by those who bear the cost of change. In the case of the Clean Air Act of 1970, the automobile companies have resisted the deadline for reaching exhaust pollution control deadlines by continual litigation, suing to block enforcement on procedural and technical grounds. Further, the auto companies have gambled on deadline extension, believing that if they did not comply, policymakers would give them more time rather than shutting them down.

One of the puzzles of the Tucson case is why water users ultimately accepted an only slightly modified water policy. Successful recall elections had no precedent in Tucson and seemed to signal politicization of the water issue and determined resistance to higher rates. How is it then that Tucsonans turned away from politics as a strategy through which to resist water rates, and instead turned toward individual response to the water use and water cost issue? The following three explanations seem to be the best.

1. Citizens became disillusioned with the political process as a means of resistance. The recall drive represented an enormous enterprise in political mobilization. Involvement was high and the election turnout was large. The performance of the new council members was a disappointment even to those intimately involved in their recruitment.[26] The new council members were clearly in over their heads. One publicly confessed not to know the party composition of the City Council. The newly elected council members had a difficult time coping with technical information about water, and they openly admitted they had been mistaken in their previous positions. They clearly had promised to roll back water rates, and they did not do so. Such a performance does not improve the citizens' image of political action as a means of accomplishment. The lesson seemed to be that the political process and politicians could not do anything about water rates, and political action imposed costs with no returns. Further, some users may have become convinced that there were physical limitations on supply that political action could not change.

2. Water users had no alternative except to pay higher rates. Municipal water departments are monopolies. Users have two alternatives: to pay their water bills or to have their supplies shut off. There is no substitute for the water service the city provides. The costs of political upheaval had been high, and the rewards in terms of reduced rates were nonexistent. Moreover, the political attention of the public and press is transient, and the water crisis inevitably was replaced by other issues. Once the recall succeeded, the motivation for outside mobilizers evaporated. There was no incentive for political entrepreneurs to develop alternatives to what the rate-making process dictated.

3. Once shocked into consciousness about water use, citizens found using less water was not too difficult. Evidence developed in chapter 5 shows that users did respond to price increases. Moreover, the response in terms of reduced water use was even greater than the estimated static demand curve would lead us to expect. There was a shift in demand which signaled that users were evaluating water in a new way and that they were adopting different water-using practices. But price remains an important variable in regulating the consumers' water use behavior. As the real price (contrasted with the nominal price) of water has fallen over the last four years, per capita water use has again begun to rise.

Conclusions

A radical change in the cost of municipal water service appears to be of low political feasibility. Services are usually on the routine, bureaucratic agenda, which means that changes are small and incremental. Yet there are circumstances under which radically different policies that broadly distribute costs can occur. The Tucson case illustrates the necessary conditions, which are as follows:

1. Innovative individuals with high levels of technical expertise within the relevant agency must initiate change and formulate the details of the policy. Thus, the vanguard of change will be municipal service managers with advanced understanding of economics and policy and with goals about resource management beyond serving customers at the least cost. Municipal officials must forgo autonomy in decision making in order to achieve larger change by offering innovative proposals to be placed on the political agenda.

2. Some local politicians must be policy leaders willing to take political risks. Among the incentives for political risk-taking are the passing of political control to a new group anxious to make a name for itself, and strong ideologies or problem-solving commitments. A few policymakers are so insulated, or else are otherwise so politically popular and powerful, that they can afford to take some unpopular actions. But programs can also be devised that bring politicians counterbalancing credit.

Someone or something must bear the political costs of unpopular action. Those who take political risks may have to suffer political setbacks and defeats, but defeat need not necessarily be the case. Action can be explained in terms of physical, economic, and financial necessity, but to be acceptable these reasons must be widely understood.

3. There must be a perception that a real problem exists. Where there are no physical indicators such as drought, establishing the existence of a problem may be difficult. Outside consultants' reports may verify the existence of a problem. Actors of different political persuasion may have to come to the same conclusion in order to convince the public that conditions are real. It can be helpful to have an "objective" nonpolitical citizens' group with technical expertise certify that there is a serious situation

that can be solved in no other way. Further, the public must believe that burdens are borne equitably. Ideology, such as the idea of conservation, could conceivably be the basis of public acceptance of costs, but the ideology must be widely shared and reasonably related to the sacrifices expected. There must be some reasonable agreement about the content of ideology, and the policy result must not seem to benefit some persons much more than others.

4. The costs that target groups are asked to bear must not be too great. There may be compensating or mediating policies that soften the impacts of costs. The policies can be symbolic as well as material. The behavior that the policy aims to modify must be relatively easy to change.

This last criterion of political feasibility of policies that broadly distributes costs leads us to a concern with practice. Some resource-using practices are more difficult to modify than others are. The next chapter discusses water-using practices, their flexibility, and the inherent possibilities for reduced water use.

Notes

1. James Q. Wilson, *Political Organization* (New York, Basic Books, 1973).

2. R. W. Cobb and C. D. Elder, *Participation in American Politics* (Baltimore, Johns Hopkins University Press, 1972) p. 14.

3. E. E. Schattschneider, *The Semi-Sovereign People* (New York, Holt, Rinehart and Winston) pp. 1–19.

4. Cobb and Elder, *Participation in American Politics*.

5. Jennifer Zamora, A. V. Kneese, and Erick Erickson, "Pricing Urban Water: Theory and Practice in Three Southwestern Cities, "*Southwestern Review* vol. 1, no. 1 (1981) pp. 89–113; and Steve H. Hanke and Robert K. Davis "Demand Management Through Responsive Pricing," *Journal of the American Water Works Association* vol. 63 (September 1971) pp. 555–560.

6. David Holtz and Scott Sebastian, eds., *Municipal Water Systems* (Bloomington, Indiana University Press, 1978) p. 3.

7. Ibid., p. 36.

8. Ibid., p. 35.

9. Steve H. Hanke, "Water Rates: An Assessment of Current Issues," *Journal of the American Water Works Association* vol. 67 (May 1975) pp. 215–219.

10. City of Tucson, *The Central Arizona Project: A Staff Report* (Tucson Water Department, 1974).

11. Interview with Stephen Davis, Tucson Water Department, Planning Division, Tucson, Ariz., October 31, 1980.

12. Interview with John Varga, 1976 Recall Campaign, Tucson, Ariz., July 23, 1980.

13. Ibid.

14. Alternatively, some congressmen spend their first terms consolidating their relationships with their districts. See, e.g., Richard Fenno, Jr., *Homestyles, House Members in Their Districts* (Boston, Little, Brown, 1978).

15. Interview with Frank Brooks, Tucson Water Department, Tucson, Ariz., November 4, 1980.

16. James Coleman, *Community Conflict* (Glencoe, Ill., Free Press, 1957) pp. 9–10.

17. Varga interview, July 23, 1980.

18. Charles O. Jones, *An Introduction to the Study of Public Policy* (Belmont, Calif., Wadsworth, 1970).

19. Interview with Mayor Lewis Murphy, Office of the Mayor, Tucson, Ariz., July 29, 1980.

20. M. R. West Marketing Research, Inc., *Water Use Study: Attitudes and Knowledge* (Tucson, Ariz., December 1976).

21. Interview with Gene Cronk, Tucson Water Department, Tucson, Ariz., November 13, 1980.

22. Interview with George Rosenberg, Citizens' Water Advisory Commission, Tucson, Ariz., July 18, 1980.

23. Murphy interview, July 29, 1980.

24. Jeffrey L. Pressman and Aaron B. Wildavsky, *Implementation* (Berkeley, University of California Press, 1973).

25. Helen Ingram and Dean Mann, *Why Policies Succeed and Fail* (Beverly Hills, Calif., Sage Publications, 1980); Robert T. Nakamura and Frank Smallwood, *The Politics of Policy Implementation* (New York, St. Martin's Press, 1980); David Mazmanian and Paul Sabatier, *Effective Policy Implementation* (Lexington, Ky., Lexington Books, 1981).

26. Varga interview, July 23, 1980

7

Practices

Water Use in Desert Cities

Cities in the deserts of the Southwest make easy targets for criticism because of the large quantities of water they use.[1] For example, residents of Phoenix, Arizona, have been cited as using about 240 gallons per capita per day, while residents of New York City use only half that much.[2] And it has been pointed out that even in Tucson, where per capita water use has been reduced significantly, per capita consumption is above the national average.[3] Such comparisons imply that people in more humid regions are careful in their use of water, while the desert dwellers are wasteful. Not addressed are the reasons for differences between the humid and the more arid regions in their use of water.

A simple fact may escape many who have not lived in a desert environment: that is, the ground will not be green unless it is watered. Grass and leafy plants require water to survive, not just to thrive. The natural desert terrain, left unmanaged by humans, is primarily rocks and dirt interspersed with often spiny desert plants. Even the most ardent desert enthusiast would probably prefer a more hospitable environment

for lounging on a lazy Sunday afternoon. People move to the southwestern desert cities for the weather and the outdoor leisure life-style.[4] The great outdoors beckons, but only after it has been turned into golf courses, swimming pools, and green landscape. The warm sunshine and dry climate that attract people to these desert cities do not produce a hospitable outdoor environment without the addition of extra water. The lawns and plants that grew so easily in the East require more care in the desert.

Houses in the desert in the summer are unbearably hot unless they are cooled, and cooling often requires large quantities of water. Homeowners increasingly are choosing evaporative cooling rather than refrigeration in order to save on high energy bills. But evaporative coolers, locally known as "swamp" coolers, require a continual flow of water. Once evaporative cooling is installed, using 150 or more gallons of cooling water a day in the summer months becomes established practice. Shade trees and green plants outside of the house also lower temperatures indoors. Staying cool and comfortable means using water.

Structural decisions to a large extent dictate water use everywhere, but perhaps especially in the desert. Researchers have found that such factors as house size and value, number of bathrooms, and number of appliances are strongly correlated with water use.[5] But in arid regions, the choice of whether or not to have a lawn and plants, a swimming pool, and an evaporative cooler is a stronger influence on day-to-day water use practices.

Behavior, of course, differs somewhat among water users who have made the same structural choices. Outdoor watering can be done every day or twice a week. The cooler can be run all night or shut off at night. Routine, more than conscious decision making, controls much of this behavior.

Habitual Nature of Water Use

Routines are an important way of dealing with the myriad of details of everyday life. Habitual behavior permits concentration on matters that require thought and conscious decision making. Patterns of water use are particularly habitual. Many of the routines of daily life revolve around the use of water—from stumbling into the bathroom first thing in the morning to wash, take showers, and flush the toilet, to making coffee in the kitchen, to washing dishes, and starting a load of clothes in the washer. Less essential uses of water also are characterized by habitual patterns of use. People tend to do their outdoor watering around the same time of the day and at regular intervals. Once established, these routines take little thought and determine a large part of water use. Desert dwellers are no more likely to consider water use consciously than are other people, and there is little reason why they should.

It would not make sense for people to take a different approach to their everyday use of water. Relative to other commodities used in the home, water is quite inexpensive. It is integral to many activities that are considered essential. The cost of water used in bathing, cooking, and drinking is insignificant compared with the values of those activities to most people. Calculations about how much water will be used and about the value of that water are not ordinarily considered at all. For residents in desert cities, using more water than is used by people living in more humid regions is a rational response to drier climatic conditions, not an irrational wasting of water. Practices involving high use of water have developed over a long period of time in desert cities such as Tucson, and they reflect routine behavior and expectations.

For policymakers interested in conserving water resources, the structural and behavioral determinants of water use practices need to be explored. What proportion of water use must be attributed to factors, such as weather, that are remote from policy influence? How much water use is related to the choice of investments in plantings, appliances, and other structures, and under what conditions are different structural decisions made? If changing behavior means changing habits, what will raise water use to the level of conscious decision in users' minds? These questions are the focus of inquiry in this chapter.

Weather—The Immutable Factor in Water Use

Water use in desert cities is closely related to climatic conditions. Use is lowest during the mild winter months and increases dramatically during the hot summer months. Indoor uses of water vary little over the course of the year. The increases are caused by seasonal outdoor use.[6] In Tucson, in 1974 and 1975, summertime pumpage (May through October) was 1.6 times the year-round average. Peak-month pumpage (usually June) was about 2.6 times the pumpage in December or January (table 7-1).

Consumption by single-family residences accounted for 64 percent of the water delivered by the city's water utility, and it is to single-family residences that the city directs its Beat the Peak campaign. Their consumption rises strongly during the spring, peaks in June, dips slightly in August because of the summer rains, and then declines gently during the fall. In the

Table 7-1. Seasonal Variation of Water Consumption in Tucson: Averages for 1974 and 1975
(million gallons per day)

| Month | Metered deliveries by class of customer | | | | | | Pumpage[a] |
	Single-family	Duplex-triplex	Multiple-family	Commercial	Industrial	Total	
January	24.52	1.08	4.27	6.93	1.80	38.58	43.9
February	26.36	1.13	4.59	7.71	2.45	42.23	49.4
March	28.28	1.11	4.51	8.16	2.61	44.65	56.0
April	35.35	1.24	4.89	9.02	3.30	53.78	70.2
May	46.96	1.45	5.63	10.89	4.96	69.87	94.4
June	65.35	1.99	7.41	14.59	7.09	96.42	113.1
July	59.09	1.99	7.22	14.28	5.97	88.54	85.6
August	48.89	1.63	6.57	12.83	6.04	75.95	97.3
September	51.58	1.80	6.99	14.19	6.01	80.55	77.7
October	37.68	1.38	5.80	11.40	4.95	61.20	63.7
November	31.10	1.29	5.43	11.40	2.92	52.83	50.5
December	25.31	1.09	4.64	8.73	2.07	43.46	44.4
Averages							
Annual	40.05	1.44	5.66	10.85	4.19	62.19	70.5
May–Oct.	51.60	1.71	6.60	13.03	5.84	78.76	88.6
Nov.–Apr.	28.49	1.16	4.72	8.66	2.53	45.92	52.4
Ratio[b]	1.81	1.47	1.40	1.50	2.30	1.72	1.69
Total (1,000 acre-feet)	44.88	1.61	6.34	12.16	4.70	69.69	79.0
	(64%)	(2%)	(9%)	(17%)	(7%)	(100%)	

Source: Tucson Water Department records, 1974 and 1975.
[a]Pumpage is shown less than deliveries in some months. Pumpage is the amount of water actually pumped in the month, while metered deliveries are taken from monthly meter readings made at different times in the month.
[b]The ratio is the May-to-October average divided by the November-to-April average.

six months from May through October (the summer-rate months), single-family residences use 80 percent more water than they do in the six winter months. Extensive garden watering, particularly lawn sprinkling, is the principal cause of this peak; its other main constituents are the use of evaporative coolers and evaporation from swimming pools.

Multiple-family residences used a further 9 percent of the water delivered by the city. The seasonal variation in use is similar in form and origin to that of single-family residences. But with their communal lawns (or no lawns) and refrigerated cooling, they are less thirsty in the summer than are detached homes—consumption by multiple-family residences in the summer was only 40 percent higher than in the winter. Duplexes and triplexes, accounting for only 2 percent of metered deliveries, have an intermediate pattern of use: here, consumption in the

summer was 47 percent more than it was in the winter.

Commercial customers used 17 percent of the water delivered by the city. Irrigation—particularly of parks and golf courses—and cooling make commercial use 50 percent more in the summer than in the winter. Customers allowed water at the reduced industrial rate accounted for 7 percent of metered deliveries. Most of this water was used by the schools of Tucson Unified School District. The major water-using industry in the city, a manufacturer of rubber gloves, used only 0.7 million gallons per day (2.9 million ft^3 per month) of the 4.2 million gallons per day used by customers qualifying for the industrial rate.[7] Because of extensive irrigation of school playing fields, industrial consumption shows more seasonal variation than does any other class of consumption—summer use was more than twice as great as winter use.

Chapter 5 showed that the evapotranspiration (EVT) rate less the amount of rain corresponds closely with the variation in monthly water use by Tucsonans. Weather conditions do not explain the decrease in average water use in Tucson since 1976, but within any given year evapotranspiration less rain does explain almost all of the variance in the average consumer's water use between months.

As has been stated, the demand for water also shows a marked cycle during each day. During the night, consumption is very low. It increases sharply in the morning to a morning peak and then declines somewhat until late afternoon. Consumption rises to an evening peak and then gradually declines to its low nighttime level. The use of water during the summer for lawn sprinkling and evaporative coolers is highest in the late afternoon and early evening, resulting in very high peak demands on summer afternoons. In Tucson the peak-hour demand was estimated to be 3.5 times the average daily demand.[8]

Evapotranspiration is the quantity of water that is both directly evaporated from plant surfaces and transpired through the plant in the growth process. EVT is affected by temperature, humidity, the latitude of the area, and the angle of the sun. EVT less the quantity of rain (EVT − R) is a close predictor of monthly water use, but temperature alone is a fair predictor. For the years 1974 through 1980, the correlation coefficient between daily pumpage and daily high temperature during the month of June averaged .61 (table 7-2). If the anomalous data

Table 7-2. Correlations Between Daily Pumpage and Daily Maximum Temperature During June, 1974–1980

Year	Correlation
1974	.75
1975	.28
1976	.64
1977	.59
1978	.61
1979	.77
1980	.65
Mean	.61

Note: Data upon which these correlations are based were provided by Tucson Water Department.

for 1975 are eliminated, the mean correlation is .67.

Indoor Use Versus Outdoor Use

Weather basically affects outdoor use. In this context, indoor use means the nonconsumptive use of water for general sanitary purposes, while outdoor use means the consumptive use of water for watering lawns and gardens, filling swimming pools, and for feeding evaporative coolers. From a water conservation perspective, saving outdoor, consumptive-use water is the important objective. Nonconsumptive water use flows into the sewer where it may be treated and used again. While it costs money to reclaim sewage effluent—and the more money the better the treatment—saving indoor water does not save water per se. Water that is used consumptively is either evaporated or transpired into the atmosphere, or absorbed into the soil without reaching a usable water table.

Much of outdoor use can affect inside temperature. Use of water for evaporative coolers directly affects the indoor environment, but the water is lost to the atmosphere. Use of water for lawns and trees also can affect the indoor environment and, in some cases, may drastically lower the residents' electrical power cooling costs. But for the sake of simplicity, we identify all uses that do not return water to the system through sewer flows as outdoor use.

The flow of sewage into the city's sewage treatment plant shows almost no seasonal variation, indicating that indoor use varies little over the course of the year. Metered deliveries are lowest in December and January, the two months when very little water is used on gardens or for cooling, suggesting that the amount of water used indoors year-round is equal to metered deliveries in January and December. Thus, any water delivered in excess of the January or December quantities in the other months of the year is used outdoors.

The divisions in consumption between indoor and outdoor use for the various classes of customers are shown in table 7-3. Because Tucson's water-consumption patterns were in a state

Table 7-3. Division in Water Consumption Between Indoor and Outdoor Use, Tucson, January and December Averages for 1974 and 1975
(million gallons per day)

Month	Single-family	Duplex-triplex	Multiple-family	Commercial	Industrial	Total
			Class of customer			
Jan.–Dec. average	24.92	1.09	4.46	7.83	1.94	40.24
	(62%)	(3%)	(11%)	(19%)	(5%)	(100%)
Differences[a]						
February	1.44	0.04	0.13	0.00	0.51	1.99
March	3.36	0.02	0.05	0.33	0.67	4.41
April	10.43	0.15	0.43	1.19	1.36	13.54
May	22.09	0.46	1.17	3.06	3.02	29.63
June	40.43	0.90	2.95	6.76	5.15	56.18
July	34.17	0.90	2.76	6.45	4.03	48.30
August	23.97	0.64	2.11	5.00	4.10	35.71
September	26.66	0.71	2.53	6.36	4.07	40.31
October	12.75	0.20	1.34	3.57	3.00	20.96
November	6.18	0.00	0.97	3.57	0.98	12.59
Average difference	18.15	0.40	1.44	3.63	2.69	26.36
Total indoor use (1,000	27.92	1.22	5.00	8.77	2.17	45.09
acre-feet/year)	(62%)	(3%)	(11%)	(19%)	(5%)	(100%)
Total outdoor use (1,000	16.88	0.37	1.34	3.38	2.50	24.52
acre feet/year)	(69%)	(2%)	(5%)	(14%)	(10%)	(100%)
Total (1,000 acre-feet/year)	44.88	1.61	6.34	12.16	4.70	69.69
	(64%)	(2%)	(9%)	(17%)	(7%)	(100%)
Ratio[b]	62:38	76:24	79:21	72:28	46:54	65:35

Note: Figures in parentheses are percent of total consumption.
Source: Tucson Water Department records, 1974 and 1975.
[a]Difference between January–December average and average for the given month.
[b]The ratio of total indoor use to total outdoor use for each class of customer.

of change after the rate increases of July 1976, these quantities are, again, based on the average of monthly consumptions in 1974 and 1975. The first row of table 7-3 shows the January-December average consumption of each class of customer. The differences between these consumptions and consumptions in the remaining ten months of the year are shown in the next ten rows. The average differences are the average of these ten differences and indicate the average amount of water used outdoors during these ten months.

The total annual indoor consumptions in table 7-3 are found by converting the January-December averages to acre-feet per year. The total annual outdoor consumptions are also converted to acre-feet per year (remember that outdoor consumption takes place only for ten months of the year). The percentages in these columns sum horizontally—for example, 11 percent of the 45,000 acre-feet used indoors is used by

multiple-family residences, and 69 percent of the 25,000 acre-feet used outdoors is used by single-family dwellings. Outdoor use totals 35 percent of all water delivered, and 69 percent of this outdoor consumption is by single-family residences. The staff of the city's Water Department suggests that evaporative coolers are responsible for about a quarter of outdoor water consumption.

Reducing Levels of Water Consumption

The water use situation described so far offers little comfort to policymakers interested in conserving water. Water-using habits tend to be routine, and low water costs do not encourage much conscious thought about use. Continued high water use offers many amenities to desert dwellers. Outdoor water use does fluctuate, but such variation is largely explained by seasons and weather. Yet, as shown in table 5-4, in chapter 5, Tucson's water pumpage per capita fell from about 190 gallons per day in fiscal year 1975/76 to about 148 gallons per day in 1978/79. Reduction in water use can occur, and water planners insist that further savings will be necessary. Our tasks are to examine how reductions can occur, what incentives can operate, what particular practices may have changed, and which are amenable to further change.

Reduction in water use may be expected to occur in three phases. Initial or short-term savings, made in response to some shock or crisis, often are limited to a change in behavior as contrasted with a change in structure. Short-term responses are simply modifications in water-using habits. Since these changes are of limited duration and are in response to real perceived problems, they may be quite extreme. Behavioral changes during the 1976–1977 California drought were three to four times more effective in reducing residential water consumption than were structural devises.[9] For example, a structural change to use of a toilet dam saved 15 percent of the water normally used, but flushing less frequently saved 50 to 75 percent of normal use. Since structural changes accounted for such a small proportion of water savings and exhortions about water use did not continue, water use rose to near normal levels after the crisis was over.[10]

Thus, practices in the medium term are likely to be less elastic than in the short term. Water use increases in the medium term as people relax their diligence in following more frugal practices. To maintain reduced water consumption would require divestment of water-using appliances or green landscape, both long-term commitments. Certainly the drastic behavioral modifications that characterize response to drought emergencies are unlikely to endure once the worst of the crisis is past.

In the long term, reduced water use is implemented through structural changes. Users become convinced that the need to cut back is here to stay, and they search for means to save that do not require daily, conscious thought. They change landscaping and purchase water-saving appliances. Because so much of water use is dictated by structure, it is only when water savings are embodied in investment choices that water savings in a city are likely to be enduring.

Data on water use in Tucson for the last three years indicate that the city has not yet achieved long-term per capita water savings (table 5-4). Behavior, responsible for a good part of the earlier reduction in use, has partially returned to old habit patterns. Individuals have made noticeable structural changes, such as removing lawns, and per capita water use is below its fiscal year 1975/76 level, but recently per capita use has again begun to rise.

A variety of techniques can be employed to encourage consumers to make short-term behavioral and ultimately long-term structural changes. The effectiveness of any measure depends upon the context in which it is introduced.[11] Since conservation techniques operate in tandem, it is a mistake to assign proportional responsibility for gallons saved to a particular measure. It would be especially misleading to suggest to other cities that some specific reduction in use will occur if they adopt a particular pricing, public information, or other strategy. However, it is possible to identify the broad requirements under which reduction in water use is likely to occur.

First of all, users must see it in their self-interest to change their water practices by using less water. Saving money is a powerful incentive, and, as was demonstrated in chapter 5, decreases in water use are closely associated with increases in water price—just as increased water use is associated with water price decreases. Preachments alone are not likely to be effective. However, while most people will not modify their practices over the long term for the common good or for ideological values, preachments can reinforce economic signals. Responding to pocketbook pinch is less painful if one thinks frugality is virtuous. Self-interest in using less water is reinforced if saving water also saves work and time (such as abandoning yard work) and conservation practices have become acceptable or even fashionable.

Second, for reductions in water use to be likely, users must become aware of their stake in changing water practices. Water use is habitual and demand is quite inelastic. Shifts in demand occur only when water users are jolted into consciously considering water use. A physical supply crisis such as was experienced in the 1976–1977 California drought may serve to sharpen awareness. As reservoir levels dropped lower and lower, the public in the affected areas saw indisputable evidence of a real problem. Once people became convinced that there was a real water supply problem, they were receptive to ideas about how to respond to the crisis.[12] The drought triggered awareness, causing people to look at water differently and to adjust their water uses.

The political events experienced in July 1976 by the people of Tucson, although not based on this kind of short-term physical supply problem, also triggered a change in how Tucsonans viewed their water resources. The immediate reaction of many people to the higher water rates and new special charges passed by the City Council was to find a political solution to the unwanted charges through a recall election. When the recall drive was successful but the new City Council kept the higher water rates, Tucsonans were forced to reevaluate the water issue itself. Having become more sensitive to the higher cost of water, Tucsonans reduced their use of water.

There is some question as to whether price increases unaccompanied by political turmoil or conservation preachments can prompt conscious awareness of water use. While response to the large 1974 price increase was masked by very hot weather, tables 5-1 and 5-4 show that water use already was declining sharply before the price increase and political furor of 1976, apparently because of price increases alone. However, in some other cities price increases alone apparently failed to result in conservation until supplemented by other actions. When East Brunswick, New Jersey, raised its water rates by 30 percent in a noncrisis environment, the upward consumption pattern continued.[13] More information is needed on the circumstances under which price alone will raise consciousness about water use. It is also important to distinguish between a nominal price increase that might

be a real price decrease in the context of an inflationary economy.

The third requirement for likely reductions in water use is that users must continually be given reinforcement about the advantages of reducing water consumption. The Tucson example shows that if the economic message of water rates is not reinforced by real rate increases, water-saving efforts will slack off. If price increases are greater than the rate of inflation, monthly bills will provide continuing signals of the need to conserve. In an inflationary era, though, the impact of the message of higher-priced water will be diminished in the din of escalating prices for all goods.

Public information programs might be an effective reinforcement technqiue. In actual practice, however, such media advertising is often a substitute for, rather than a reinforcement of, other policies. Information programs provide policymakers with noncontroversial and relatively inexpensive means to try to influence the public in its use of water. Because these campaigns generally are aimed at encouraging voluntary conservation measures and often are not tied to pricing or regulatory strategies, they are acceptable and widely used.[14] A few studies claim success in the reduction of water use because of educational campaigns.[15] However, as Baumann points out, in these cases educational efforts were combined with a variety of other techniques, and the effect that the educational campaigns themselves exerted is unknown.[16]

The major educational program intended to affect Tucsonans' water practices has been the Beat the Peak campaign. Residents have been exhorted never to water their lawns and gardens between 4:00 P.M. and 8:00 P.M., to water not more than every other day, and never to water on Wednesdays. This campaign was not designed to lower total water use, but simply to spread the use out over the day, lowering use during the peak hours. If the plan worked, less capacity would be needed in the distribution system, and money (not water) would be conserved.

The campaign has been declared a great success by the Tucson Water Department.[17] But there is some reason to believe that the success in beating the peak has been only coincidental with the decline in per capita water use and not a direct effect of Beat the Peak rhetoric.

The regression analysis presented in table 4-3 and figure 4-4 in chapter 4 showed that the ratio of peak monthly use to average annual monthly use remained constant over all levels of use at approximately 2 from 1974 through 1979. This period includes years both before and after Beat the Peak was instituted. Peak monthly use has fallen, but only in direct proportion to the decline in total per capita use. If the campaign had affected peak use independently of total use, the ratio would have changed after 1976, showing different ratios at different levels of mean monthly use. The decline in per capita use can largely be explained by price increases, combined with a shift in demand related to the political events. Thus, if Beat the Peak has had any effect at all, it has been an indirect effect through reinforcing public awareness of water prices. Again, we recognize that our test is with monthly data rather than with daily or hourly data. Still the reduction in the daily or hourly peak could be fully explained by the reduction in use, and it is unlikely that the monthly ratios would remain so constant if peak hourly use were merely being distributed more evenly throughout the day.

The Role of Attitudes and Perceptions

The relationship of attitudes to conservation behavior is complex and poorly understood. While researchers believe that attitudes are important, there is little consensus about how important they are or about the way in which they have impact.[18] Certainly attitudes would seem to be less significant than factors such as weather and prices. Lupsha, Schlegel, and Anderson concluded:

Overall, it would appear that the impact of attitudes on demand is a relatively weak and minor aspect of any water use equation. While attitudes do have an impact, it is small and tends to reflect self interest, which probably has stronger surrogates in economic and appliance variables and the micro-environmental factors which are probably better tested by the landscape and sprinkling practice variables.[19]

Bruvold, studying effectiveness of conservation programs in drought-stricken California, identifies consumers' perceptions of the seriousness of the situation as a key attitudinal factor.[20] Consumers who have been taught to think differently about the provision of water service in their community are more likely to adjust their use of the resource. But the teaching requires people to reevaluate their beliefs about water as it affects them personally. Residents of a drought-stricken community may come to fear that their water supply will be cut off if they do not respond with reductions in use. Tucsonans also experienced a sense of vulnerability in their use of water. Rate hikes and new special charges made consumers reevaluate their beliefs about inexpensive and abundant water at their own faucets. Tucsonans responded to this new sense of cost and risk by reducing use.

The immediacy of these perceived risks should be distinguished from more general beliefs about water scarcity. There is little dispute that a substantial proportion of the public believes that water shortages are a potential problem. For example, in a public opinion survey conducted in Tucson in 1976, 71 percent of the people said that they believed water shortages were a serious or very serious problem (table 7-4).

These opinions were expressed at a time when per capita water use was almost 190 gallons per day. Later that same year, and after the political storm over water rates had risen, another sample

Table 7-4. Perceived Seriousness of Water Shortages by Tucson Residents, 1976

Perceived seriousness	Percent of respondents
Very serious	26.0
Serious	45.2
Not sure	18.5
Not very serious	7.5
No problem	2.7
	100.0

Source: Survey reported in Helen Ingram, Nancy Laney, and John McCain, *A Policy Approach to Political Representation: Lessons from the Four Corners States* (Baltimore, Johns Hopkins University Press for Resources for the Future, 1980).

of Tucsonans was asked if there was a water problem in the Tucson area. Only 60 percent of the Tucsonans polled said that water was a problem.[21] This same question was asked in two subsequent surveys in 1980 and 1981, and the proportion of people perceiving a serious water problem in Tucson remained quite constant.[22] The proportion of people undecided about whether there was a problem decreased in the subsequent polls, but the proportion of Tucsonans who believed there was *not* a problem increased. Since substantial reductions in water use occurred during this same period, the linkage between the attitudes and the water-using practices of Tucsonans is uncertain.

Common pool resources like water present difficult problems for people in terms of integrating their beliefs and practices. People may sincerely believe that water is scarce and that conservation of the resource is important. But unless there are personal benefits in reducing one's own water use, it is difficult to justify sacrifices for a nebulous general good. Without appropriate pricing or regulatory signals, there is no reason to believe that personal sacrifice by one consumer is not lost to increased waste by another. The contradiction lies not in disparities between consumer belief and practices, but rather in public policy that treats water as a common pool resource without limits.

The utility of attitudinal data to those interested in changing water-using practices is circumscribed by these and other considerations. Regulations and public information campaigns can be geared to address the equity problems inherent in water as a common pool resource. As made clear in chapter 6, people's attitudes about water determine in large part the political risks faced by those who would change existing water policy. But, the fact that price increases are an unpopular strategy for reducing water use does not mean that price therefore is ineffective as a conservation tool. Lord, Chase, and Winterfield note that there may be an inverse relationship between the effectiveness of conservation techniques and public acceptability as measured in opinion surveys.[23] The existence of such a relationship would be accepted a priori by most economists. After all, what individual would

want an action that would tend to diminish his or her short-run welfare? However, public attitudes about price as a conservation tool must be part of the calculus that politicians consider in deciding upon conservation strategies.

As more research is done in the area of values and behavior, the relationships between expressed attitudinal factors and consumer practices may begin to be revealed. In the meantime, one must recognize that the connection between consumers' responses to public opinion polls about water and their own water-using practices is at best indirect.

Changing Outdoor Practices

The present demand for water is in part the result of consumers' tastes and preferences, in part the result of their lack of knowledge about how to use less water, in part habitual behavior, and in large part structural. The demand for water used outdoors is largely the result of a preference for green lawns. Although some reductions in outdoor use can be made by watering more carefully, by installing covers on swimming pools, and by keeping evaporative coolers in good repair, significant reductions in outdoor use can be made only by the structural measure of reducing the amount of greenery in parks and gardens.

One alternative to actually reducing greenery that has been widely advocated in Tucson is the use of sewage effluent on parks and golf courses and on the agricultural lands surrounding the city. However, because this action does not reduce greenery, it does not, in fact, conserve water.

In Tucson, the quantity of sewage effluent discharged from the treatment plants is about 90 percent of the water used indoors. Currently, about 60 percent of the city's sewage effluent infiltrates the bed of the Santa Cruz River between the treatment plants and the Rillito Narrows just north of the city, and an additional 30 percent infiltrates between the Rillito Narrows and the county line another 15 miles north. About 54 percent of the water used indoors, then, returns to the Tucson Basin's stock of water, and another 17 percent of the water used indoors replenishes the groundwater of northern Avra Valley from which the city could also pump.[24] If the effluent is sold to farms or the mines, or is given to the Papago Indians in order to settle the present dispute over the basin's water, this recharge will not take place and the water will be used consumptively. Similarly, watering gardens, parks, or golf courses with wastewater will not conserve water, since this use merely consumes water that otherwise would have returned to the ground.

Another alternative for reducing outdoor use would be to switch from evaporative cooling to refrigeration. Evaporative coolers require considerably more water on site than do refrigeration units. These coolers use fans to force air through wet-pad mediums, which causes water to evaporate, raising relative humidity and lowering air temperature. Water consumption by this type of cooler is quite variable depending upon weather conditions (temperature, relative humidity), the water recycling system within the cooler (total, partial, or no reuse), length of daily operation, and size of the cooler. It has been estimated that on hot, dry summer days, coolers may use an average of 200 gallons per day.[25] A more realistic whole-season daily consumption use estimate would be about 125 gallons.[26] Such use represents an average addition of 30 to 40 gallons to the daily per capita water use in these households. Evaporative coolers are generally chosen over refrigeration because of their lower energy requirements (about 85 percent less), and therefore their water consumption is not likely to be an important factor in the decision to stay with or change from this type of cooling.

In fact, the trend in recent years, since the advent of high energy prices, has been toward the use of evaporative coolers and away from refrigeration. In 1976 more than 80 percent of Tucson homes had evaporative coolers, although some also had a refrigeration unit to use during the very hot and humid weeks of late summer.[27]

Of course from a systemwide perspective there may be little difference in water consumption between the two types of cooling. Evaporative coolers use more water on site, while refrigeration units use more water in the off-site production of electricity needed to run them. In any case, it is not likely that much water saving could, or will, come from measures associated with cooling, at least in the near future.[28]

Swimming pools are another major structural determinant of outdoor water use among Tucsonans. The water requirements to maintain a swimming pool and a grass lawn of equal size are about equivalent.[29] In a 1979 sample of Tucson water users, slightly more than 10 percent had pools.[30] Correlations between summer water use by the people in the sample and their having or not having a swimming pool were significant, although quite low, ranging from .14 to .19, depending on the year. Pools do not explain a large part of overall water use, but they do represent a stable determinant of outdoor water use for those people who have them. People may be willing to get rid of their green lawns to save water, money, and trouble, but they are not likely to abandon their pools for those reasons. Swimming pools represent a substantial capital investment and are more closely associated with the owners' life-style values than are lawns.

Thus, greenery in parks and gardens becomes the crucial issue in reducing outdoor water use. Traditionally, Tucson landscaping practices had reflected the preference of migrants from more humid areas of the country for green lawns and lush vegetation. It is estimated that green lawns use about 123 gallons per week per 100 square feet (ft^2) of lawn.[31] Averaged over a year, the water requirement for a grass lawn accounts for nearly 30 percent of the total average household consumption.[32] But a 1979 study revealed that beginning in 1976 a number of Tucsonans have removed their lawns and thereby reduced their water use (table 7-5).[33] The study found that there had been a 2.8 percent increase in irrigated backyard lawns and a 0.5 percent increase in front yard lawns among their sample of Tucson homes from 1972 through 1975. In 1976 that trend was reversed, and from 1976 to 1979 there was a 14.8 percent decrease in irrigated backyard lawns and a 20.3 percent decrease in front yard lawns (table 7-5). The trend was more pronounced for front lawns than for back lawns, indicating that the social conformity aspects of lawn ownership are less important than the functional enjoyment criteria.

As early as 1975 it had been hypothesized that landscaping tastes were changing in Tucson with the resurgence of appreciation for the desert environment.[34] The existence of this attitude was mostly among the residents residing at the periphery of the city, in the foothills areas, but it was speculated that the trend toward appreciation of native vegetation had led to increasing use of desert landscaping in established neighborhoods. These avant-garde (and fairly wealthy) residents may have provided a new model in acceptable fashion to the preference of migrants from more humid areas for green lawn landscaping. It also may even be that recent immigrants to Tucson are more amenable to desert landscaping than are longer-term residents. A recent survey found that a higher percentage of people who had lived in Tucson for more than ten years said green lawns were "very important" to them, than did residents who had lived there fewer than five years.[35]

The change to desert landscaping also has been encouraged by its ease of maintenance. Unlike green lawn and tree landscaping, which requires frequent watering, mowing, and pruning, desert landscaping can be quite carefree, particularly where the conversion is accomplished by just letting the grass die, rather than by expensive relandscaping using native vegetation. Apparently the lower recreational and aesthetic value of the yard is compensated for by the time and effort saved in not having to maintain a green environment. In any case, the "browning" of Tucson is tending to occur.

Table 7-5. Percentage of Sample Households Removing Their Green Lawns, Tucson, 1972–1979

Period	Front yard	Backyard
1972–1975	− 0.5	− 2.8
1976–1979	20.3	14.8

Source: David A. Mouat and Michael Parton, "Assessing the Impact of the Tucson Peak Water Demand Reduction Effort on Residential Lawn Use, 1976–1979," Office of Arid Lands, University of Arizona, December, 1979.

Changing Indoor Practices

It has been estimated that about three-quarters of all water used indoors is used in the bathroom for flushing, washing, and bathing. The standard toilet uses about 5 gallons of water per flush, or 25 gallons per capita per day. A shower or bath usually takes about 20 gallons of water. Kitchen use accounts for only about 3 gallons per capita per day, while laundry can consume another 50 gallons per load (table 7-6).

In a study of Albuquerque water users, it was found that the number of water-using appliances in a household was significantly related to consumption.[36] Households with fewer than seven such appliances used fewer than 345 gallons daily, while those with more than twelve used more than 580 gallons per day. Even higher levels of consumption were found for those people owning "luxury" appliances, including dishwashers, garbage disposals, humidifiers, and ice-making machines. As luxury items become more commonplace, especially in new housing, higher indoor water use is structurally predetermined.

A countertrend may be found in the "new technology" of conservation. A wide array of devices to save water is now available. Many manufacturers of plumbing supplies and water-using appliances have added a line of products with lower water requirements. The reason that these devices are not widely used relates not to their availability or cost, but simply to the preferences or ignorance of consumers. A 1979 study of household demand for water in Indiana found that despite the availability of a wide variety of water-saving devices and appliances, very few people were aware of them.[37] The researchers estimated low economic returns to a household from the purchase of these items and hypothesized that this low return explained people's lack of motivation to seek out information on these devices.

However, once consumers become receptive to the idea of reducing their water use, the attraction of these devices could be enhanced considerably if their water savings are linked to energy savings. Table 7-7 shows some of the more common water-saving fixtures available,

Table 7-6. Indoor Water Use per Person per Day

Fixture	Gal/use	Use/day	Gal/day	Use rate
Toilet	5	5	25.0	5 gal/flush
Bathing	20	1	20.0	4 gal/min
Lavatory	—	—	2.0	—
Dishwasher	15	0.25	3.75	—
Kitchen sink	—	—	3.0	—
Laundry	50	0.18	8.75	50 gal/load
Utility	—	—	1.25	
Gal/person/day			63.75	

Source: Larry K. Baker. "Experience and Benefits of the Application of Minimum Flow Water Conservation Hardware." in *Proceedings of the National Water Conservation Conference on Publicly Supplied Potable Water.* U.S. Department of Commerce. National Bureau of Standards publication 624 (Washington. D.C., June 1982) p. 282.

their range of likely water use reduction, their original cost to the consumer, and their potential for energy savings. The double incentives of higher water prices and increased energy costs for heating that water could induce a trend toward consumer acceptance and adoption of the "new technology."

For example, a water-saving clothes washer costs about $25 more than a standard one, but would save the average household about $2 each month in water and fuel costs; a faucet aerator costing $2 would save $0.32 each month; a water-saving toilet costing little more than a standard one would save about $0.90 a month.[38]

A simple device that would pay for itself in a month or two is a dual-flush mechanism. By using such a device, either a large flush of 5 gallons or a small flush of 1.5 gallons may be selected according to the user's requirements. In table 7-6 the breakdown of indoor water use

Table 7-7. Potential Water Savings from Residential Interior Fixtures

Fixture/action	Water use		Percent savings (gal/day)		Incremental original cost ($)		Energy savings
	Standard	Improved	New	Retro[a]	New	Retro	
Tank toilet	5–7 gal per flush	3.5 gal per flush	10–18	10–18	$0–$10	$0–$6	No
Shower	Up to 12 gal/min	3.0 gal/min	9–12	9–12[b]	$0–$5	$1–$5	Yes
Kitchen and lavatory faucets	Up to 5 gal/min	1.5 gal/min	0–2[c]	0–2[c]	$0–$5	$1–$5	Yes
Pressure-reducing value	80 lb/in²	50 lb/in²	0–10	0–10	$0–$25	$25	Yes
Automatic clothes washer[d]	27–54 gal per load	16–19 gal per load	0–5	—	$20–$30	Not practical	Yes
Automatic dishwasher	7.5–16 gal per load	7.5 gal per load	0–4[c]	—	0	Not practical	Yes
Total	—	—	19–51[f]	19–42[f]	—	—	—

Source: State of California. The Resources Agency. "Water Conservation in California." Department of Water Resources Bulletin No. 198. May 1976.

[a]Retrofitting may not always be practical.

[b]Attachments marketed with 0.5 gal/min flow. Residential acceptance unknown. but commercially proven.

[c]No field qualification.

[d]Of the households in Los Angeles area. 59 percent have washing machines and 24 percent have dishwashers.

[e]Based on one load per day.

[f]Education to wash only full loads. turn off water faucets unless actually used. etc.. could add another percent or 2 to totals.

shows that dual-flush toilets would cut the amount of water used for flushing by half, reducing indoor water consumption by 20 percent—this would cut monthly water consumption of an average family of four by about 220 ft^3. If the marginal price of water averaged $1.00 per 100 ft^3 over the year, $2.20 would be saved each month by installing a dual-flush mechanism. So-belov and Lloyd report that the tenants of an apartment complex in Birmingham, England, were perfectly satisfied with dual-flush toilets installed on an experimental basis.[39] The whole-sale price of these devices is $1.75, but they are not available in Tucson.[40]

This discussion suggests that the demand for water used indoors results from ignorance as well as from preference. While outdoor use can be reduced significantly only by reducing the popularity of green lawns, a campaign advising the public of ways to save both water and money could reduce indoor water use, making total water use less resistant to change than is suggested by the demand function estimated in chapter 5. But, as discussed earlier, saving indoor, *noncon-sumptive-use* water really saves very little water at all. It would save money for water consumers.

Some Future Alternatives

In fiscal year 1977/78, as a result of real price increases, water problem awareness caused by the political furor of 1976, and the continuing Beat the Peak campaign, water consumption in Tucson fell to a low of 147.5 gallons per capita per day. Since that time, real water prices have fallen, and although the Beat the Peak campaign continues and water is a subject in the news almost every day, per capita daily use has risen to 166 gallons.

There is, however, continued interest on the part of public agencies, especially the Arizona Department of Water Resources, in reducing per capita use. The 1980 Groundwater Management Act has given them a mandate to do so. Thus, Tucson will surely be the focus of future schemes aimed at reducing consumption.

Any future scheme is certain to be a combi-nation of changes in the water rate structure, the launching of a continuing public education campaign, and the introduction of regulations restricting water use or requiring the installation of water-saving devices. The effect of any future scheme aimed at reducing water consumption is unpredictable, particularly in view of the un-certainty of effect of an educational campaign (which would be a central part of any scheme, making both increases in the price of water and restrictions on the use of water more palatable). Nevertheless, we posit two alternative possible levels of reduced consumption and their effects on the Tucson environment.

It would be reasonably easy to achieve a per capita use of 130 gallons per day. This level assumes that outdoor use would be reduced to 35 gallons per person per day, and per capita daily indoor use would be 95 gallons. There appears to be a trend toward townhouses and desert landscaping in Tucson, so only modest reductions in the area of lawns and a little more care in watering would be required to bring con-sumption down to this level. Even this level of reduction, however, would require significant price increases if price alone were to be the motivating force. Given the demand elasticity estimated in chapter 5, a real price of about $1.11 per 100 ft^3 in 1979 prices would be re-quired in the mean-use block. The current nom-inal price is only about $0.98, and the real price in 1979 dollars is less than $0.70. Thus the current price would have to be raised by almost 50 percent to about $1.45 per 100 ft^3 in the mean-use block if significant education cam-paigns or legal measures were not used to shift the demand curve to the left. Coincidentally, such a percentage price increase is in the lower end of the range suggested in chapter 4 if the water utility were to attempt to switch to mar-ginal cost pricing.

If per capita use falls to 130 gallons per day, and if the Tucson population increases at what is forecast as a "medium" rate, total pumpage would need to rise to about 96.5 million gallons per day by the year 2000, up from fiscal year 1980/81 pumpages of 69.7 million gallons per day.[41]

It would be much more difficult to achieve a per capita daily water use of 85 gallons. Such a level would be required if total pumpage in 2000 were to be about at current levels, given a "medium growth" population increase. Under this level of consumption, 70 gallons and 15 gallons per day per capita are assumed for indoor use and outdoor use, respectively. Such a reduction in outdoor use would require a very severe reduction in greenery in Tucson. Since an average of 8 gallons per person per day is used over the year by evaporative coolers, little water would be left over for watering lawns—desert landscaping would be the norm, and parks, playing fields, and golf courses would be small and few. The city would indeed be brown. The water price that would be required in order to achieve such reduced use certainly would be high. Our available data do not allow an estimate.

Conclusions

Changing water use practices in a desert city through public policy is possible, but far from easy. High per capita use in arid land communities does not mean that use is more frivolous than it is elsewhere and can be cut back without difficulty. There are good reasons for using lots of water in the desert, and the benefits of high water use far outweigh current costs. Further, many of the determinants of water use are beyond the manipulation of public policy. Data cited in this chapter and in chapter 5 show that demand for water is closely associated with weather. Further, levels of use are strongly determined by long-term structural decisions about investments in appliances, swimming pools, cooling, and landscaping that take time and strong incentives to modify. Moreover, much of water-use behavior is routine and is not easily raised to the conscious level at which consumers can make knowing decisions about change. However, as the Tucson case illustrates, given the proper circumstances, changes in practices can occur.

Most important among incentives is the conscious realization among consumers that it is in their self-interest to lower water use. Rate increases provide the incentive, especially when they are accompanied by events making users sensitive to the issues. Lowered use will remain low only when consumer behavior is reinforced by price signals and public information. Long-term savings will be made only as structural changes are made that inherently use less water.

This chapter identified a number of ways in which water use can be lowered, particularly in Tucson. Whether or not these practices should be adopted is a matter of applying one's values in weighing benefits and costs. Obviously, the overdraft on the groundwater in the Tucson Basin cannot continue forever. Yet there are at least 10 million recoverable acre-feet of water in the basin. Whether this resource should continue to be mined and at what rate are matters of choice. The ultimate need for water conservation exists. But Tucsonans surely should decide just how much they should encourage their city to grow, and just how brown they think a desert city should be.

Notes

1. For example, an editorial in *The Washington Post*, October 12, 1979, entitled "Water Shortage Everywhere," emphasized widespread waste of water by pointing out that "Arizona, with only 10 inches of annual rainfall, ranks among the top 10 states in per-capita water consumption." (Cited in U.S. Environmental Protection Agency, *Residential Water Conservation: An Annotated Bibliography*. FRD-16 (Washington, D.C., February 6, 1980).

2. The 1981 National Public Radio Journal Broadcast, "At The Last Watering Hole" (Washington, D.C., National Public Radio, 1981).

3. Dunn and Larson reported an average *household* water demand of 148 gallons per day for a community with no restrictions on watering. Dorothy F. Dunn and Thurston E. Larson, "Relationships of Domestic Water Use to Assessed Valuation with Selected Demographic and Socioeconomic Variables," *Journal of the American Water Works Association* vol. 55 (April 1963) pp. 441–450.

4. Neal B. Pierce, *The Mountain States of America* (New York, Norton, 1972).

5. P. A. Lupsha, D. P. Schlegel, and R. O. Anderson, *Rain Dance Doesn't Work Here Anymore* (Albuquerque, University of New Mexico Division of Governmental Research, 1976); and G. A. Watkins, "A Sociological Perspective of Water Consumers in South Florida Households," *Florida Water Resources Research Center* vol. 18 (1972).

6. Charles W. Howe and F. P. Linaweaver, Jr., "The Impact of Price on Residential Water Demand and Its Relation

to System Design and Price Structure," *Water Resources Research* vol. 3, no. 1 (1967) pp. 13–32.

7. City of Tucson Water Department, unpublished data, 1979.

8. John Carollo Engineers, "Report Submitted to the Metropolitan Utilities Management Agency," Final report (Tucson, Ariz., March 11, 1976).

9. Mark Hoffman, Robert Glickstein, and Stuart Liroff, "Urban Drought in the San Francisco Bay Area: A Study of Institutional and Social Resiliency," *Journal of the American Water Works Association* vol. 71 (July 1979) pp. 356–363.

10. William H. Bruvold, "Consumer Response to Urban Drought in Central California," Final report, USF Grant no. ENV 77-16171 (Washington, D.C., Research Applied to National Needs, June 1978).

11. D. D. Baumann, "Information and Consumer Adoption of Water Conservation Measures," in *Proceedings of the National Water Conservation Conference on Publicly Supplied Potable Water*. Publication 624 (Washington, D.C., U.S. Department of Commerce, National Bureau of Standards, June 1982) pp. 179–190.

12. Mark Hoffman, Robert Glickstein, and Stuart Liroff, "The Impact of Large Temporary Rate Changes on Residential Water Use," *Journal of the American Water Works Association* vol. 71 (July 1979); and Richard A. Berk, C. J. LaCivita, Katherine Sredl, and Thomas F. Cooley, *Water Shortage* (Cambridge, Mass., Abt Books, 1981).

13. L. Mason Neely, Michael J. Opaleski, Theodore B. Shelton, and Dennis Palmini, "Conservation in a Non-Crisis Environment—Township of East Brunswick, New Jersey," in *Proceedings of the National Water Conservation Conference on Publicly Supplied Potable Water*. Publication 624 (Washington, D.C., U.S. Department of Commerce, National Bureau of Standards, June 1982) pp. 433–441.

14. For examples of such campaigns, see A. P. Brigham, "Public Education Campaign to Conserve Water," *Journal of the American Water Works Association* vol. 68 (December 1976) pp. 665–668; Century Research Corporation, *Social Aspects of Urban Water Conservation*. Prepared for the Office of Water Resources Research, August 1972 (available from National Technical Information Service, Springfield, Va. PB 214970); and Berk and coauthors, *Water Shortage*.

15. Arthur P. Brigham, "Water Conservation and the Drought," *Water and Sewage Works* vol. 65 (July 1977); and Donald G. Larkin, "The Economics of Water Conservation," *Journal of the American Water Works Association* vol. 70, no. 9 (1978) pp. 470–474.

16. Baumann, "Information and Consumer Adoption of Measures."

17. Gene E. Cronk, "Results of a Peak Management Plan for Tucson," in *Proceedings of the National Water Conservation Conference on Publicly Supplied Potable Water*. Publication 624 (Washington, D.C., U.S. Department of Commerce, National Bureau of Standards, June 1982) pp. 453–464.

18. Baumann, "Information and Consumer Adoption of Measures."

19. Lupsha, Schlegel, and Anderson, *Rain Dance* p. 23.

20. Bruvold, "Consumer Response."

21. M. R. West Marketing Research Inc., *Water Use Study: Attitudes and Knowledge* (Tucson, Ariz., December 1976).

22. Judith Dworkin, "Identifying Public Preferences for Water Conservation Measures." Paper presented at the Applied Geography Conference, Tempe, Ariz., October 22, 1981.

23. William B. Lord, James A. Chase, and Laura A. Winterfield, *Evaluation of Demand Management Policies for Conserving Water in Urban Outdoor Residential Uses*. Research Report 82-1 (Boulder, Colo., Policy Sciences Associates, January 1982).

24. Adrian H. Griffin, "An Economic and Institutional Assessment of the Water Problem Facing the Tucson Basin" (Ph.D. dissertation, University of Arizona, 1980) p. 133. Currently the city has wells in the more southern portions of the Avra Valley.

25. Jeffrey Cooke, "Cool Houses for Desert Suburbs" (Phoenix, Arizona Solar Energy Commission, June 1979).

26. K. James DeCook, "Water Conservation for Domestic Uses," prepared for the City of Tucson by the University of Arizona (1977).

27. M. R. West Marketing Research Inc., *Water Use Study*. In the more humid Phoenix area, only 25 percent of the households have evaporative cooling.

28. Should the development of photovoltaic cells make home-grown electricity from the sun cost-effective, a reverse trend of reverting to refrigeration could begin.

29. DeCook, "Water Conservation for Domestic Uses."

30. David A. Mouat and Michael Parton, "Assessing the Impact of the Tucson Peak Water Demand Reduction Effort on Residential Lawn Use, 1976–1979 (Tucson, Office of Arid Lands, University of Arizona, December 1979).

31. DeCook, "Water Conservation for Domestic Uses."

32. Mouat and Parton, "Assessing the Impact."

33. Ibid.

34. Melvin E. Hecht, "The Decline of the Grass Lawn Tradition in Tucson," *Landscape* vol. 19, no. 3 (May 1975) pp. 3–10.

35. Mouat and Parton, "Assessing the Impact."

36. Lupsha, Schlegel, and Anderson, *Rain Dance*.

37. Rodney L. Clouser and William L. Miller, *Household Demand for Water and Policies to Encourage Conservation* (West Lafayette, Ind., Purdue University Water Resources Research Center, 1979).

38. U.S. Army Corps of Engineers, "Water Quality Planning Report submitted to the Pima Association of Governments." Final report (Tucson, Ariz., U.S. Army Corps of Engineers, Los Angeles District, PAG-208 Project Office, June 1978).

39. A. Sobelov and C. J. Lloyd, "Trials of Dual-flush Cisterns," *Journal of the Institution of Water Engineers* vol. 18, no. 1 (February 1964) pp. 53–58.

40. Duo-Flush of Colorado Springs, Colorado. Advertising Material for Duo-Flush Devices. 1979.

41. Griffin, "An Economic and Institutional Assessment," pp. 140–142.

8

Lessons from the Tucson Experience

Retrospection in Policy Analysis

Policy analysis discovers how policy intent becomes embodied in actual practice. The role of the analyst is to link policy purpose to policy actions, and policy actions to real-world consequences. There is a striking need for such analysis of water conservation policy. Water scarcity is a popular topic with the media, and convincing people to use less water is the conventionally prescribed remedy. Further, water conservation techniques are the focus of a great deal of contemporary scholarly research. This literature identifies a number of actions that authors believe are necessary to bring the quantity of water demanded and the available supply into balance. Yet people continue to use a great deal of water despite the concern with scarcity expressed in the media, and outside of conditions of periodic drought, there are few policy actions taken to lower the quantities demanded. The link between the intention of saving water and the reality of saving water has not been forged.

Since successful implementation of policy objectives is the aim of policy analysis, the analysis needs to be rooted in actual experience.

The disjunctive and disorderly world of politics with its unpredicted events and unanticipated consequences is beyond the analysts' power to simulate without reference to real-world examples. Herein lies the attraction of the Tucson experience. Between 1974 and 1979 per capita water use in this desert city dropped by 28 percent. Our task as policy analysts has been to explain how and why this saving occurred and to draw what lessons we could for the practice of water conservation in this and other cities.

The Tucson experience was not the successful achievement of some preset conservation goal. On the contrary, the history of the experience indicates that policymakers did not agree on policy intent and did not anticipate what actually occurred. The observer is forced to conclude that had the Tucson Water Department staff and the members of the City Council known where they were going in 1976, they would have chosen a more direct and less hazardous route. Nor is the city currently pursuing a policy that clarifies intent and rationally relates aims to policy actions that will modify water-using behavior. It is reasonable to question whether the city's water policy is really water conservation, and there is evidence that per capita use has begun to climb rather than to decline.

The historical roots of conservation are so deeply conflictual that, despite conservation's symbolic appeal, it has been difficult to make the concept operational. As demonstrated in chapter 3, identifying conservation with use and development leads to situations in which the costs of water supplies, taking into account political, social, and environmental as well as economic costs, may be far greater than are their benefits to users. It is no more justifiable to define conservation as nonuse, assuming that the less water used the better. If the benefits of particular water use exceed the costs of supplies, and no other more valuable present or future water uses are forgone by that particular use, then saving water makes no sense. The Tucson case suggests that water conservation should comprehend both water supply and water demand management and should employ analysis of benefits and costs in the choice of appropriate strategies.

All too often, the real test and purpose of policy—accomplishing policy aims—is lost by policy analysts in prescribing procedures. There are various recipes or sets of rational rules about how policy decisions ought to be made. Often these rules are related to disciplinary perspectives. The philosopher has guidelines for formulating coherent, consistent concepts. The economist uses certain analytical tools to arrive at efficient solutions to problems. The political scientist is preoccupied with the strictures that govern the electoral process and the conditions of obtaining and holding political power. Physical scientists focus upon physical relationships and the physical forces that must be set in motion or altered to deliver the desired physical response. Retrospective analysis of the Tucson experience dramatically illustrates that following any one set of rational rules defies the precepts of other disciplinary orientations that are equally important to successful policy impact.

The concepts of conservation espoused in Tucson were contradictory; thus the rationale provided for action was confused. The Tucson Water Department and most other policy actors saw conservation as a means to financial well-being and orderly development. The members of council majority in 1976 were attracted to conservation as a component of preservation doctrine. They envisioned nonuse of water, not further development. In terms of logic, the concept of conservation lacked satisfactory definition. Yet, confusion may be politically functional. A coalition was forged between politicians and professionals in favor of innovative municipal water rates in the name of conservation. Further, the preachment of conservation, confounded though it is, has helped to justify in water users' minds their reduction in water use. In the long term, however, we would expect that the lack of an orderly definition will have adverse political consequences. Policymakers and the public are likely to discover that the persons asked to make sacrifices in the name of conservation—the water users—and those who reap the benefits—the Water Department, developers, and builders—are not necessarily the same people, and there will be objections. Yet if the concept of conservation is to be helpful in achieving policy aims, the political utility of the preachment must be preserved while it is made more logical and consistent.

The Tucson experience also illustrates the stress that exists between economic rationality and political feasibility in municipal water ratemaking. Marginal-cost pricing, the most efficient form of pricing from an economic perspective, means substantial price increases in situations where supplies are becoming more expensive. Yet under ordinary circumstances it is not politically rational to recommend large rate increases. Under some conditions, however, it is possible to resolve at least partially the conflict between economics and politics. Taking political risks and bearing political costs is reasonable for political actors who have substantial political resources and for whom there are countervailing ideological and problem-solving rewards. Although not as common as we might wish, policy leadership is now and then practiced. Further, there may be means whereby economically rational rates may be made more politically palatable—by phasing in rate increases or by making compensation for inequitable or extremely adverse impacts of rate increases.

What is advantageous politically may not provide the necessary ingredients to change people's water-using practices. But the political furor that accompanied the 1976 rate increases, while fatal to a number of political careers, made Tucson water consumers more sensitive to water issues. Routine water-related activities were suddenly given conscious attention, and some were changed. In drought situations, physical water shortage and the absence of rain trigger increased consciousness about water. The Tucson case shows that a "crisis" can be created and still be effective in modifying users' behavior. While shocks are always likely to be politically dangerous for the public officials held responsible for their delivery, shocks that make the public more sensitive to surrounding events can be politically useful. By illustrating what did not work politically, the Tucson experience provides some guidance toward safer political paths. Had the public relations campaign that accompanied Beat the Peak begun at the time of the price increase, users' first response might not have been political. Instead of demanding the recall, users might have more quickly become reconciled to altering their water-using behavior. The price increase might have been politically palatable if the Citizens Water Advisory Council that was later appointed had lent its impartiality and expertise in certifying the real nature of the water problem before action was taken.

Unconventional Conclusions

An important role of policy analysis is to draw lessons from the past that reduce the possibility of future error. Study of the Tucson case throws doubts on some false assumptions and conventional wisdom about water policies that still exist and need to be questioned, as the six following points show.

First, it is not true that public approval of conservation techniques is necessary to their successful use. Opinion survey findings indicating that people prefer voluntary conservation programs to price increases or regulations ought not to be taken at face value. It is understandable

that people would prefer programs under which burdens are voluntary. Noncoercive public information campaigns also make sense to politicians. They make few enemies, and allow public officials to take credit for whatever water savings occur. It is also reasonable to suppose that, outside of drought conditions, strictly voluntary programs will not be much of a constraint on individual behavior. In a commons situation it is not sensible for any user to cut back as long as he or she cannot be certain that nearly all other users are doing the same. Otherwise the user makes an individual sacrifice without perceptibly alleviating the shortage problem. Effectiveness is as important a criterion as acceptability, and it is unlikely that voluntary programs by themselves will be effective. The Tucson case shows that people will accept price increases, and they will respond by cutting water use. Municipal utilities are monopolies, people have no substitutes for water, and in the end water users have little alternative to paying the going rates.

Second, political disruption is not necessarily a sign of water policy failure. Conflict is expensive in political terms, and political actors spend a great deal of time and energy avoiding and containing conflict. There is a difference, however, between the experience of political problems and the failure to implement policy intent. In the Tucson case, the political turmoil resulting from the 1976 rate increase made people sensitive to the water issue and triggered changed behavior. While the council majority members responsible for the innovative policy were removed from office, their policy has endured and has had substantial impact. In our interviews with Tucson Water Department staff, current public officials, and interest-group leaders, the 1976 council members often were spoken of as failures and the days of upheaval in Tucson were seen as bad times. Current water officials speak of restoring the department's damaged image.[1] As indicated in chapter 6, we believe that, had the New Democrats been more sensitive to matters of political feasibility, they might well have saved their jobs. Yet in a policy sense, it may be that it is impossible to make an omelet without breaking a few eggs. Controversy was a key ingredient in saving water in Tucson.

Third, it is not true that water conservation is always beneficial to environmental quality. One of the tenets of preservation-oriented environmentalists is that self-denial with respect to use of natural resources is virtuous and that nonuse of a resource is bound to make the environment better off. Yet, as illustrated in chapter 3, a variety of actions benefiting quite diverse interests are taken in the name of conservation. In the case of water conservation in Tucson, the sacrifices asked for in the Beat the Peak campaign facilitate population growth and housing development. While an aesthetic argument can be made that desert landscaping is appropriate to an arid environment, it is a strange twist of environmental values to condemn green plants to die of thirst and to prefer hot summer months without the blessing of shade trees. Environmentalists are frequently accused of preferring landscape over people. Conserving water in Tucson facilitates a different result. Many more people using less water will be able to live in the Tucson Basin.

The Tucson case is an example of a lesson already taught by a number of authors—conservation is a complex and contradictory concept with great symbolic appeal that is employed by various groups for their own benefit.[2] There is no a priori reason for supposing that conservation serves environmental values more than it serves any other values. Conservation is not an end in itself, but rather a means to achieve some aim. It is reasonable to question the purposes of saving water and to ask whether benefits exceed costs.

Fourth, increasing block rates, despite their appeal to environmentalists, do not really reflect the costs of service. The real costs of water service depend on the cost of delivery of an additional unit of water, the costs of providing additional units of peak-day and peak-hour capacity, and the cost of providing for an additional customer.

Since there is never a capacity problem in the winter months, the only costs are that of the water itself and that of providing service. Since each unit of water contributes equally to the additional water cost, there is no rationale for an increasing block rate schedule in the winter other than a general desire to hold down total water use regardless of its costs or benefits. This desire reflects the preservationist point of view.

Peak-load costs occur only in one or two summer months. Even then, since all customers using water at the time of the peaks contribute to the peaks, an increasing block rate will not reflect the real costs of delivering water unless the larger water users' peaking ratios (the ratio between peak use and average use) are higher than those for a smaller water user. Only in this case will the larger water users' contribution to peak demand on the system increase more when an additional unit of water is used than will the small water users' contribution. The evidence in Tucson indicates that the peaking ratios are constant between users of small and large quantities of water.

Fifth, while water professionals tend to regard water use as water needs, demand for water is not totally inelastic, and amounts used are related to rates charged. After we satisfy our physiological need for bodily fluids, there is no one quantity of water that is absolutely required. All evidence in Tucson and elsewhere indicates that while the demand for municipal water is relatively inelastic, it is not totally inelastic. Therefore, any change in the real price of water will have a small but not zero effect on per capita water use. Many people find this assertion counterintuitive, since they say they have no idea what the price of water is. Enough people to make a difference must have some idea, however, because the historical record documents price response.

It is important to distinguish the real price of water from the nominal price. The real price recognizes the effects of inflation and indicates the changes in the price of water relative to the changes in the price of other commodities. In the past several years in Tucson, the nominal price of water has risen each year, but the real price has fallen. Thus, per capita water use has risen.

Sixth, popular notions to the contrary, urban residents of arid regions are not more likely than are urban dwellers in humid regions to waste water. Media commentaries on water problems in the West frequently cast aspersions upon desert citizens for using more water than do residents of New York or other cities.[3] If, however, waste means that costs to consumers are disproportionate to value received, arid land dwellers may be less wasteful. Water is priced low everywhere, but in desert cities the benefits of water use are often large. The desert untreated by water is not particularly hospitable. Treated with water, however, barren soil can become an inviting oasis. Data cited in chapters 5 and 7 indicate that water use in Tucson is highly correlated to evapotranspiration and weather. Arid land dwellers use more water because the climate is hotter and drier than in the humid Northeast, not because they are more profligate.

More water is used for outdoor watering in desert cities than in cities in wetter climates where indoor use is a higher proportion of total use. This fact does affect the flexibility that the two different populations have in cutting back water use. Conservation in desert cities means losing outdoor amenities important to general life-style. Conservation in wet areas means changing household habits of convenience, comfort, and cleanliness.

Lessons for Municipal Water Policy

In chapter 1, we noted that Tucson's water problems are not different in kind from those faced by many other municipal water utilities. Ever-increasing demands have resulted from higher living standards and urbanization. The geographical area of water supply for many cities is often small, and new reservoir sites are scarce. Groundwater supplies are limited and sometimes not of necessary quality. Interbasin transfers from long distances are expensive and environmentally damaging. Even though new supplies are difficult and costly, water is underpriced in most cities, thus encouraging use. Dealing with these problems will require innovative municipal water management, yet policies that are effectively designed and implemented to deal with water problems are

rare. Chapter 1 suggests that meaningful change in municipal water management can occur through the meshing of the four P's: preachment, price, politics, and practice. Altering any of the P's is difficult, and accomplishing changes in one area that are consistent with the other dimensions is even more difficult. The Tucson experience offers the following insights about how such coordinated change might come about.

Economic reasoning is likely to have its greatest impact upon municipal water management through communication with the staffs of municipal water utilities. The experience in Tucson indicates that water professionals have the largest stake in innovative water policy and are most likely to be initiators of changes. In the past these professionals have not been particularly responsive to economic thinking. Their concerns have been with physical efficiency and with covering the total costs of capital improvements. However, the growing problems faced by water managers have caused them to search for new ideas. Because the Tucson Water Department was in financial difficulty, it was open to considering cost-of-service concepts. While the increasing block rates and other changes adopted did not represent sound economic thinking, nevertheless, Tucson Water Department staff indicated a willingness to use price to manage demand.

According to Hanke, much of the blame for failure to influence municipal water policy lies with economists themselves.[4] They have not provided sufficient concrete examples of how to apply their economic theories, nor have they spoken in language or through channels to which water managers are receptive. Chapter 4 in this volume addresses this communication gap with specific criticisms of the guidelines for rate setting and with concrete suggestions for changes.

Techniques for water saving designed by engineers and other professionals are more likely to be adopted and to have the desired impact when they are consistent with and responsive to economic and other incentives for change. The technicians who construct water-saving strategies are preoccupied with impacts on physical systems and all too often disregard economic and social appeal. It does little good to manufacture water-saving appliances for which there is no market. Only when it can be shown that such appliances have benefits that outweigh the costs—perhaps in energy as well as water savings—are such appliances likely to be purchased. But while important, price is not always the only or even the most important reason for adopting changes. As pointed out in chapter 7, native vegetation around houses was first adopted by wealthy newcomers. Switching to desert landscaping in Tucson was fashionable, and cactus requires less upkeep.

Goals of environmental quality are more likely to be achieved when the choice of conservation strategies is guided by benefit-cost analysis. Conservation, as it is often narrowly construed and practiced, is not always good for the environment. Instead, a more integrated concept of conservation, comprehending water demand and supply management, is more desirable. The role of economic analysis in such an integrated concept is crucial, since the choice of whether to develop additional supplies or to manage demand depends upon relative benefits and costs, taking fully into account environmental costs.

There are increasing indications that an integrated conception of conservation that leans heavily upon benefit-cost analysis now has a more receptive audience, particularly among some environmentalists. At one time environmentalists encouraged everyone to act out their respect for the natural world and to husband resources without any demonstrated need. However, after more than a decade in which environmental values have been widely adopted, mere symbolic actions aimed at gaining public attention no longer make sense. Many environmentalists are now in search of ways to implement their environmental goals in the form of effective policies. Thus, they are more open to analysis and quantification.

Water utility managers will be more successful in legitimizing innovative water policies when they recognize the risks their proposals present to elected politicians. Water utility managers have been content in the past with the dentist's theory of utility management—delivering technically competent, painless service. However,

the problems faced by contemporary utilities can no longer be handled in the conventional bureaucratic, valuefree manner. Important choices must be made among overall goals, and it must be decided which groups are to bear what burdens concerning water service. Such decisions will involve political decision making and political risk.

Water managers must learn to operate in a political environment. They need to become more sensitive to the political valence of different water management strategies. They must communicate in terms that part-time, lay public officials understand. They will have to help marshal political support for their proposals. As the Tucson case illustrates, citizens' advisory committees can be a useful political device. Headway toward making utility proposals legitimate can be made by gaining the support of such bodies.

Lessons for Tucson

It is an axiom of historians that people who fail to learn from the past are bound to repeat it. A number of signs suggest that the trend in Tucson is not toward more manageable water problems. Substantial overdrafts of the water table continue. Population is projected to rise sharply. Increases in water rates have not kept pace with inflation, and per capita water use has been climbing steadily for the past few years.

The City of Tucson currently is facing mounting pressures. The Arizona Groundwater Management Act of 1980 requires that "safe yield" be achieved in coming years. Reaching that goal will demand the adoption of additional water conservation measures. The Tucson aqueduct of the Central Arizona Project (CAP) promises to deliver new supplies of surface water to the city. However, the federal government expects a large state and local contribution toward the costs of construction. It is planned that city users will subsidize farm users of CAP water. Consequently, the CAP will be very expensive for municipal users.

We have recommended a series of actions for the city. We believe that the commodity of groundwater itself has a value and should have a cost. This cost should be in addition to the costs of water transport and delivery. A sensible way to begin to measure this value is by replacement cost. The financial burden anticipated in bringing in CAP water provides an appropriate yardstick. It is sensible policy to charge more for water now, since water is going to be a great deal more expensive when the CAP comes on line. Otherwise Tucson citizens will be receiving inaccurate messages about the true value of water, and will commit themselves to using more water than they would if they knew the eventual cost. The pressures currently facing Tucson also provide an opportunity to move toward more accurate cost-of-service pricing. The increasing block rates currently charged have some ideological appeal, but no convincing economic rationale. Eventually they may generate troublesome perceptions of inequity among water users.

Our advice with regard to Tucson water policy would require substantial and perhaps costly changes if implemented. Yet the most profound lessons of the events in Tucson are that the policy system is more flexible than it appears, policy leadership can become available, and change is possible.

Notes

1. Interview with Gene Cronk, Tucson Water Department, Tucson, Ariz., November 13, 1980.

2. Orris C. Herfindahl, "What is Conservation?" in D. L. Thompson, ed., *Politics, Policy and Natural Resources* (New York, Free Press, 1972); Leonard Shabman, "Water Conservation as a Basis for Reforming Water Resource Programs." Paper presented at the American Water Resources Association Meetings, Las Vegas, Nevada, September 24–27, 1979; and Dean Mann, "Institutional Framework for Agricultural Water Conservation and Reallocation in the West: A Policy Analysis," in Gary D. Weatherford, ed., *Water and Agriculture in the Western U.S.: Conservation, Reallocation, and Markets* (Boulder, Colo., Westview Press, 1982).

3. For example, The 1981 National Public Radio Journal Broadcast, "At the Last Watering Hole" (Washington, D.C., National Public Radio, 1981).

4. Steve H. Hanke, "Pricing as a Conservative Tool: An Economist's Dream Come True?" in David Holtz and Scott Sebastian, eds., *Municipal Water Systems* (Bloomington, Indiana University Press, 1978) p. 238.

Appendix

Estimation of Demand Where Price is Expressed as a Multiple Block Rate Schedule

Concepts

The City of Tucson Water Department charges its customers according to a multiple-part tariff in which the price of a unit of water increases with a customer's monthly consumption. For cases in which a consumer faces a price schedule rather than a fixed price, Taylor developed a model, later modified by Nordin, in which the amount of water purchased is expressed as a function of the marginal price faced by the consumer and a second quantity defined as the difference between the consumer's bill as determined from the rate schedule and the product of the marginal price and the amount of water purchased.[1] Taylor's and Nordin's analyses were confined to a theoretical consideration of a consumer's behavior and did not include any empirical studies of demand.

Basing their demand study on Taylor's and Nordin's work, Billings and Agthe made a regression analysis, attempting to explain the variation in mean monthly water consumption in Tucson by the variation in marginal price,

difference, income, and weather.[2] Griffin and Martin criticized that study as follows.

In cases when the price paid by a consumer varies with his consumption in accordance with a price schedule, the results of a conventional regression analysis will be a biased estimate of the demand function. Although Taylor's and Nordin's model of consumer behavior correctly specifies marginal price and difference as the quantities determining the amounts of water purchased by a consumer facing a multiple-part tariff, the demand function will not be estimated correctly by a regression analysis using marginal price and difference as explanatory variables. The regression analysis will be affected by the form of the rate schedule with the result that the relationship between price and quantity indicated by the regression will not be a demand function, but will be a relationship resulting from the combined effect of the rate schedule and the actual demand function.

The rate schedule will affect the results of a regression analysis because the observations of price and difference are not the prices and differences corresponding to the points at which the demand curve intersects the rate schedule; rather, the observed prices and differences are the prices and differences corresponding to the observed consumptions which, because of the error term in the regression model, lie elsewhere along the rate schedule. Consequently, there is a tendency for observed consumption to vary with observed price and observed difference in a manner quite independent of any influence that the marginal price of water or the implicit tax or subsidy from the intramarginal part of the rate schedule may have on the amount of water purchased.[3]

Griffin and Martin concluded that the extent of bias introduced into the results of the regression model would depend on the variance of the disturbance term. When the variance of the disturbance term is small, the observations will be close to the demand curve, mostly lying in the same use block, so the bias will be small. But as the variance of the disturbance term increases, more and more of the observations will lie in the inner and outer use blocks and the bias will increase.

For the case of an *increasing* block rate tariff, as the variance of the disturbance term increases, more and more of the observations will lie in the inner and outer use blocks where different marginal prices apply: the higher marginal prices in the outer use blocks and the lower marginal prices in the inner use blocks will tend to rotate the regression line clockwise from the actual demand curve so that demand will appear to be less elastic than it really is. When the variance of the disturbance term becomes very large, the regression line will be rotated past the vertical, eventually tending to run along the rate schedule, thus indicating the supply curve to the consumer rather than the demand curve.

The case with difference is similar. As the variance of the disturbance term increases, more and more of the observations will lie in the inner and outer use blocks; the high values of observed difference in the inner block and the low values of observed difference in the outer blocks will tend to rotate the regression line counterclockwise from the real demand curve toward the rate schedule. Thus, the regression model will tend to exaggerate the effect of difference on the amount of water purchased.

In order to estimate the demand function correctly, the price corresponding to the point at which the demand curve intersects each rate schedule must be determined. It is these prices,

not the prices corresponding to the observed consumptions, that must be used in the regression. Consider the situation shown in figure A-1. During the study period, three different price schedules are in force; five observations are made while each of these schedules is in force, making a total of fifteen observations. The demand curve would be estimated correctly by using the observed consumptions as the dependent variable, and the prices corresponding to the use block intersected by the demand curve as the explanatory variable. This set of observations of price and quantity is shown in figure A-1 by the circles. When the first price schedule is in force, the correct price would be 40¢ for all five observations, even though two of the observed consumptions fall in the second use block where the price is 35¢, and one of the observed consumptions falls in the fourth use block where the price is 45¢. Similarly, the correct price for observations made when the second rate schedule is in force is 60¢, and the correct price for observations made when the third schedule is in force is the price for the second use block of this schedule, which is 75¢. The use of the prices corresponding to the actual consumptions—giving the observations of price and quantity marked by crosses—will not estimate the demand function correctly.

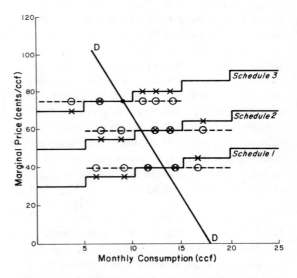

Figure A-1. Demand curve for a consumer facing several multiple-part tariffs.

The selection of the correct price to be used in the regression analysis obviously presents a problem. The correct price is the price in the use block intersected by the demand curve—but the use blocks in which the demand curve intersects the price schedules are not known until the demand curve has been estimated. For example, there is no way of knowing that the demand curve shown in figure A-1 intersects the first two price schedules in the third use block and the third price schedule in the second use block unless the demand curve has already been estimated. This seemingly intractable problem can be resolved by making a series of approximations to the demand, using the prices in the use blocks intersected by each approximation in a new regression analysis to find a new approximation to the demand curve. In this way, the actual demand curve is approached by a series of iterations.

The iterative process is as follows. Suppose the demand function is $Q = A + BP$. First a regression is performed using the prices in the use blocks intersected by the line $Q = \bar{q}$, where $\bar{q}$ is the mean of the observed consumptions. The estimated intercept A and regression coefficient B are then used as a first approximation to the demand curve.

A second regression is performed using the prices in the use blocks intersected by the first approximation to the demand curve. The intercept and regression coefficient found from this second regression are used as a second approximation to the demand curve, and the prices in the use blocks intersected by this demand curve are used to perform a further regression. The process is continued until there is no further change in the estimated values of A and B.

This method is easily adopted for a demand analysis using several explanatory variables. For example, a combined time-series-and-cross-section analysis might use the following regression equation:

$$Q_{ij} = A + BY_i + CW_j + DP_{ij} + U_{ij}$$

where Q_{ij} is the quantity purchased by the ith customer in the jth month, Y_i is the ith customer's income, W_j is the weather in the month j,

P_{ij} is the price in the use block intersected by the customer's demand curve, and U_{ij} is the disturbance term. Figure A-2 shows this situation. The price corresponding to the observed consumption Q_{ij} is not the correct price, because the disturbance term carries the observation into a different use block from the one in which the demand curve for that particular Y_i and W_j intersects the rate schedule.

The method is applied by first making a regression analysis attempting to explain the variation in consumption by the variation in all of the explanatory variables except price. In this case, the initial regression equation would be

$$Q_{ij} = A + BY_i + CW_j + U_{ij}$$

The values of A, B, and C found from this initial regression are used to find

$$\overline{Q}_{ij} = A + BY_i + CW_j$$

for each residence i for each month j. This equation is used to make a first estimate of the marginal price faced by each household in each month. The price is the price in the use block intersected by the line $Q = \overline{Q}_{ij}$ as shown in the

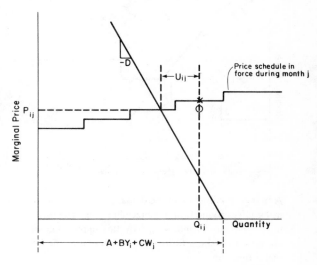

Figure A-2. A combined time-series-and-cross-sectional analysis.

uppermost part of figure A-3. The prices found from this step are used along with the rest of the data in a regression attempting to explain the variation in consumption by the variation in all of the explanatory variables; that is:

$$Q_{ij} = A + BY_i + CW_j + DP_{ij}$$

Another regression is then made, using the prices in the use blocks intersected by this demand function. The lower part of figure A-3 shows this situation. The new estimated values of A, B, C, and D found from this regression are used to find a new set of prices to be used in a further

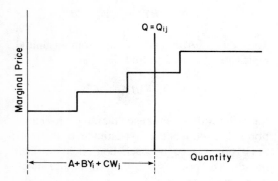

Results of initial regression are used to make a first estimate of marginal price.

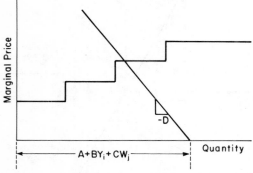

A first approximation of the demand function is made using prices found from the preceding step. The prices in the use blocks intersected by this approximation are used to make a further approximation of the demand function.

Figure A-3. The iterative regression procedure.

regression. The procedure is repeated until successive regressions show no changes in coefficients.

A complex computer program is necessary to handle this iterative procedure. It would be necessary to integrate a similar iterative procedure into the program if the effects of difference were to be included in the analysis. Since difference is such a small part of a consumer's bill, we neglect difference effects in our empirical analysis.

Results

The results of the regression on forty-two months of data for 2,159 households are displayed in table A-1. Coefficients stabilize after five iterations.

Using the large forty-two month times 2,159 observations per month data base, the estimated coefficients on the price variable do not change drastically between iteration 1 and iteration 5. However, when using a small 160-household subgroup of the random sample during testing of the program, much different results were obtained. The coefficient on price in the first iteration was $-.0032$, but stabilized at $-.0149$ after nine iterations. The price elasticity of this subgroup was much more elastic than for the random sample of the total Tucson population.

Notes

1. Lester D. Taylor, "The Demand for Electricity: A Survey," *Bell Journal of Economics* vol. 6 (Spring 1975) pp. 74–110. Taylor's model was modified by John H. Nordin, "A Proposed Modification of Taylor's Demand Analysis: Comment," *Bell Journal of Economics* vol 7 (Autumn 1976) pp. 719–721.

2. R. Bruce Billings and Donald E. Agthe, "Price Elasticities for Water: A Case of Increasing Block Rates," *Land Economics* vol. 56 (February 1980) pp. 73–84.

3. Adrian H. Griffin and William E. Martin, "Price Elasticities for Water: A Case of Increasing Block Rates: Comment," *Land Economics* vol. 57 (May 1981) p. 269.

Table A-1. Regression of $\log_{10} Q = a + bP + cEVT\text{-}R + dFCV$

Iteration number	Estimated coefficients				R^2
	Intercept	Price	Evapotranspiration minus rain ($EVT\text{-}R$)	Full cash value	
0	.729527	0	.036498	.000005	.101
1	.812536	−.003412	.040116	.000005	.102
2	.820169	−.003727	.040447	.000005	.102
3	.820691	−.003748	.040469	.000005	.102
4	.820733	−.003750	.040471	.000005	.102
5	.820737	−.003750	.040471	.000005	.102

Note: Means are

$\log_{10} Q = 0.984$

$EVT\text{-}R = 3.22$ inches

Full cash value = \$30,400

Mean for price in iteration 5 = 29.6¢/100 ft^3

(Prices are in 1966 values.)

Index

ABOUT THE AUTHORS

William E. Martin has engaged in research and teaching at the University of Arizona since receiving his doctorate in agricultural economics at the University of California, Berkeley in 1961. Previous books and articles have focused on agricultural water issues, the public lands, and farm labor.

Helen M. Ingram is a Professor of Political Science at the University of Arizona and has published books and articles on natural resource and environmental policy. She is a graduate of Oberlin College and Columbia University.

Nancy K. Laney has a law degree, a master's degree in public administration, and has taken graduate training in the Hydrology and Water Resources Administration Department at the University of Arizona.

Adrian H. Griffin earned his Ph.D. in water resources administration at the University of Arizona in 1980. He is now with the California Department of Water Resources.